THE NEXT INTELLIGENCE REVOLUTION

Why Artificial Intelligence Will Redefine Work,

Power, and the Future of Civilization

Saint Blanc

ISBN 979-8234-03101-3
First Edition

Imprint: Imprint InExt Home
Published in the United States of America

The views expressed in this book are those of the author and are intended for informational purposes only. While every effort has been made to ensure accuracy, the author makes no representations or warranties regarding the completeness or accuracy of the information contained herein.
For permissions, inquiries, or additional information, contact:
—Saint Blanc

The Intelligence Revolution Series

Book 1 — The *Next Intelligence Revolution*

Book 2 — Economy *(Coming Soon)*

Book 3 — Systems *(Future Release)*

Book 4 — Conflict *(Future Release)*

Book 5 — Future *(Future Release)*

An ongoing exploration of the forces shaping the future of intelligence, technology, and civilization.

The Saint Blanc Doctrine

The world changes the moment someone refuses to accept the impossible.

Every breakthrough in history began with a person who saw beyond the limits others accepted.

—Saint Blanc

Table of Contents

AUTHOR'S NOTE..I

AUTHOR'S PREFACE...II

INTRODUCTION...IV

PROLOGUE..IX

PART I..1

THE RISE OF MACHINE INTELLIGENCE..1

CHAPTER 1..7

THE THRESHOLD..7

CHAPTER 2..16

THE GLOBAL AI ARMS RACE..16

CHAPTER 3..27

THE INFRASTRUCTURE BEHIND INTELLIGENCE....................................27

PART II...37

THE EXPANSION OF ARTIFICIAL INTELLIGENCE..............................37

CHAPTER 4..40

THE RISE OF ARTIFICIAL INTELLIGENCE AGENTS..............................40

CHAPTER 5..49

THE TRANSFORMATION OF WORK..49

CHAPTER 6..60

THE ARTIFICIAL INTELLIGENCE ECONOMIC EXPLOSION.....................60

PART III...68

THE INTELLIGENCE ECONOMY..68

CHAPTER 7..71

THE ARTIFICIAL INTELLIGENCE STARTUP EXPLOSION.......................71

CHAPTER 8..80

THE INTELLIGENCE ECONOMY FRAMEWORK......................................80

CHAPTER 9..90

THE ARTIFICIAL INTELLIGENCE WEALTH MAP...................................90

CHAPTER 10..100

THE $100 TRILLION AI RACE...100

PART IV...112

INDUSTRIES BEING REWRITTEN...112

CHAPTER 11...115
ARTIFICIAL INTELLIGENCE IN MEDICINE AND DRUG DISCOVERY............................115
CHAPTER 12...124
ARTIFICIAL INTELLIGENCE AND THE FUTURE OF FINANCE...................................124
CHAPTER 13...133
ROBOTICS AND THE PHYSICAL AI REVOLUTION...133

PART V...**143**

THE FRONTIER OF INTELLIGENCE...**143**

CHAPTER 14...147
THE POWER BEHIND ARTIFICIAL INTELLIGENCE...147
CHAPTER 15...157
THE PATH TOWARD ARTIFICIAL GENERAL INTELLIGENCE....................................157
CHAPTER 16...166
THE ETHICS OF ARTIFICIAL INTELLIGENCE...166
CHAPTER 17...174
ARTIFICIAL INTELLIGENCE AND GLOBAL POWER...174

PART VI..**182**

THE WORLD AHEAD...**182**

CHAPTER 18...185
THE FUTURE ECONOMY IN AN AI WORLD...185
CHAPTER 19...194
HUMAN AI COLLABORATION...194
CHAPTER 20...203
THE 2040 INTELLIGENCE ECONOMY...203
CHAPTER 21...214
DESIGNING THE INTELLIGENCE ECONOMY..214
CHAPTER 22...222
THE ECONOMICS OF INTELLIGENCE..222
CHAPTER 23...229
FIVE FORCES DRIVING THE INTELLIGENCE REVOLUTION......................................229

PART VII...**236**

THE BIG QUESTIONS..**236**

CHAPTER 24...238
THE INTELLIGENCE REVOLUTION MANIFESTO...238
CHAPTER 25...244

WHAT THIS MEANS FOR YOU...244
CHAPTER 26..248
BIG IDEAS FROM THE INTELLIGENCE REVOLUTION...............................248
CHAPTER 27..253
A FRAMEWORK FOR READING AI BEYOND THIS BOOK.........................253
A FINAL THOUGHT..257
EPILOGUE...258
ACKNOWLEDGMENTS...263
REFERENCES AND RESEARCH SOURCES...264
ABOUT THE AUTHOR...268

Author's Note

AI is a rapidly evolving field, and many of the technologies discussed in this book continue to develop at an extraordinary pace.

The purpose of this work is not to provide a technical manual for building artificial intelligence systems. Instead, it offers a broad exploration of the technological, economic, and societal forces shaping the rise of machine intelligence.

Throughout the book, the term artificial intelligence refers broadly to systems capable of performing tasks that typically require human cognitive abilities. These include technologies such as machine learning models, neural networks, natural language processing systems, and intelligent automation tools.

Because the field evolves quickly, some of the developments described here may advance significantly after publication. New discoveries, improved algorithms, and emerging applications may expand the capabilities of artificial intelligence beyond what is currently known.

Where possible, this book focuses on underlying trends rather than short-term technological details. These trends include the growth of computing power, the increasing availability of data, the rise of machine learning infrastructure, and the global competition to develop advanced AI systems.

The goal of this book is to help readers understand not only how artificial intelligence works, but also why it matters.

AI is not simply another technological innovation.

It may represent one of the most significant transformations in the history of human progress.

Author's Preface

AI is often described as one of the most important technological developments of our time. In the span of just a few years, systems capable of generating text, analyzing images, writing software, and assisting with complex problem-solving have moved from research laboratories into the daily lives of millions of people.

For many, this transformation has felt sudden.

Yet the roots of artificial intelligence stretch back decades. The ideas that power modern AI, including neural networks, machine learning, and computational theory, were first explored by scientists long before the technologies required to fully realize them existed. What has changed in recent years is the convergence of several powerful forces: massive datasets, unprecedented computing power, and breakthroughs in algorithm design.

Together, these elements have accelerated the development of machine intelligence at a pace few anticipated.

This book was written to explore where that acceleration may lead.

Rather than focusing solely on the technical details of artificial intelligence, the goal of this work is to examine the broader landscape that surrounds it. AI is not merely a field of computer science. It is rapidly becoming an economic engine, a geopolitical factor, and a transformative force across industries ranging from medicine and finance to transportation and scientific research.

Understanding AI therefore requires a multidisciplinary perspective.

Throughout these chapters, the reader will encounter discussions about infrastructure, economics, ethics, global competition, and the evolving relationship between humans and intelligent machines. These subjects are deeply interconnected, and together they form the foundation of the AI revolution now unfolding.

It is important to acknowledge that predicting the future of technology is inherently uncertain. Many of the developments described in this book remain speculative. Artificial intelligence continues to evolve, and new breakthroughs may reshape the landscape in ways that are difficult to foresee.

However, by examining current trends and technological trajectories, we can begin to understand the possibilities that lie ahead.

Artificial intelligence may accelerate scientific discovery, increase economic productivity, and help address complex global challenges. At the same time, it raises questions about governance, employment, privacy, and the responsible development of powerful technologies.

The future of AI will not be determined by algorithms alone. It will be shaped by the decisions made by researchers, entrepreneurs, policymakers, and citizens around the world.

This book is intended as an exploration of that future.

Whether you are an entrepreneur, a student, a technology enthusiast, or simply someone curious about the forces shaping the coming decades, the hope is that these pages will provide insight into one of the most consequential technological transformations in human history.

The age of artificial intelligence is still in its early chapters.

What we choose to build, and how we choose to guide it, will determine the story that follows.

Saint Blanc

Introduction

The Beginning of a New Intelligence Era

Throughout human history, progress has often been defined by the tools we create.

The discovery of agriculture allowed civilizations to form. The industrial revolution introduced machines capable of performing physical labor at scales never before imagined. The digital revolution connected billions of people through global networks of information.

Each of these technological shifts transformed the structure of society.

Today, humanity may be entering another transformation, one centered not on physical tools, but on intelligence itself.

For most of the history of life on Earth, intelligence belonged exclusively to biological organisms. The ability to reason, learn, and adapt evolved slowly over millions of years through natural selection.

Humans eventually became the most cognitively advanced species on the planet. With language, mathematics, and scientific reasoning, humanity built civilizations capable of exploring space, decoding the genetic code, and constructing complex global economies.

Yet even at the height of technological progress, intelligence remained fundamentally biological.

That assumption is beginning to change.

Artificial intelligence systems are now capable of performing tasks that were once considered uniquely human. Machines can recognize images, translate languages, generate written content, assist with scientific research, and analyze complex datasets.

These capabilities are expanding rapidly.

Advances in computing power, machine learning algorithms, and large-scale data analysis have accelerated the development of artificial intelligence at a pace few experts predicted.

What once seemed like distant science fiction is becoming part of everyday life.

Students use AI systems to assist with studying and research. Entrepreneurs rely on AI tools to design products and analyze markets. Scientists are exploring how machine learning might accelerate discoveries in medicine, physics, and materials science.

AI is no longer confined to research laboratories.

It has entered the global economy.

But the rise of machine intelligence raises profound questions.

How will AI reshape industries and labor markets?

Which nations and companies will lead the development of intelligent technologies?

What new opportunities, and risks, will emerge as machines become increasingly capable?

Most importantly:

What will the world look like as artificial intelligence continues to evolve?

This book explores those questions.

Rather than focusing solely on technical details, the chapters that follow examine the broader landscape of artificial intelligence.

You will encounter discussions about the infrastructure powering modern AI systems, the economic forces driving global investment, the industries being transformed by machine intelligence, and the ethical challenges societies must address as these technologies grow more powerful.

AI is not simply another technological trend.

It may represent the emergence of a new form of intelligence on Earth.

And like every major technological revolution before it, the consequences will extend far beyond the laboratories where the technology was first developed.

The decisions made during the early years of the AI era, by researchers, entrepreneurs, policymakers, and citizens, will shape the trajectory of this technology for decades to come.

We are still in the early chapters of the artificial intelligence revolution.

But the story is already unfolding.

The pages that follow explore where this revolution began, how it is reshaping the world today, and where it may lead in the future.

The next intelligence revolution has already begun.

The question is not whether it will transform our world.

The question is how.

The Central Thesis of This Book

Throughout history, the technologies that reshaped civilization shared a common characteristic: they dramatically expanded human capability.

Agriculture allowed societies to support larger populations. Industrial machinery amplified physical labor. Digital computers accelerated the processing of information.

Artificial intelligence represents the next stage of this progression.

For the first time at global scale, intelligence itself is becoming a scalable technology.

Machine learning systems can analyze vast datasets, generate insights, and assist with complex problem-solving at speeds far beyond human capacity. As these systems become more capable, they may dramatically increase the productive potential of individuals, organizations, and entire economies.

This book explores a central idea:

AI is not simply another technological innovation. It is the foundation of a new economic and technological system: the intelligence economy.

The Intelligence Economy is an economic system in which the most valuable resource is no longer raw materials, physical labor, or even information itself. Instead, the most valuable resource becomes intelligence, meaning the ability to analyze complex systems, generate insights, automate decision-making, and solve problems at scale.

Artificial intelligence expands access to this resource by allowing intelligence to be embedded within software systems capable of operating globally. As a result, individuals, organizations, and entire nations may gain economic advantage based on how effectively they develop and deploy intelligent technologies.

In this system, the most valuable resource is no longer raw materials, physical labor, or even information itself.

The most valuable resource is intelligence: the ability to analyze complex systems, generate new ideas, and make effective decisions.

Artificial intelligence expands access to this resource.

As a result, the intelligence revolution may reshape industries, accelerate scientific discovery, and redefine the structure of global economic power.

The chapters that follow explore how this transformation is unfolding, and what it may mean for the future of human civilization.

How to Read This Book

AI is a rapidly evolving field that touches many different areas of technology, economics, science, and society.

Because of this, the chapters in this book are designed to explore the development of AI from several perspectives.

Some chapters focus on the technological foundations of artificial intelligence, explaining how advances in computing power, machine learning models, and data infrastructure have enabled recent breakthroughs.

Other chapters examine the economic and industrial impact of AI, exploring how machine intelligence is transforming industries such as healthcare, finance, robotics, and software development.

Later chapters turn toward the future, discussing emerging technologies such as AI agents, the global competition for AI leadership, and the long-term implications of machine intelligence for society.

Readers may approach this book in several ways.

Those new to artificial intelligence may find it helpful to read the chapters in order, beginning with the early sections that explain the rise of machine intelligence and the technological foundations behind modern AI systems.

Readers already familiar with the technical background may choose to focus first on the later chapters, which explore the broader implications of artificial intelligence for the global economy and the future of work.

Each chapter is written to stand on its own while also contributing to a larger narrative about the transformation currently unfolding.

AI is still in its early stages, and the technologies discussed throughout this book continue to evolve. Some of the developments described here are already reshaping industries today, while others remain emerging possibilities that may unfold in the years ahead.

The goal of this book is not simply to explain how artificial intelligence works.

It is to explore what the rise of machine intelligence may mean for the future of human progress.

Prologue

The Intelligence Threshold

For most of human history, intelligence belonged exclusively to biological life.

From the earliest organisms to modern humans, the ability to think, learn, and adapt emerged from the intricate networks of neurons inside living brains. Intelligence evolved slowly through millions of years of natural selection.

Humans eventually became the most cognitively advanced species on Earth. With language, reasoning, and creativity, humanity developed science, built civilizations, and created technologies capable of reshaping the planet itself.

Yet even at the height of technological progress, intelligence remained fundamentally biological.

Until recently.

At the beginning of the twenty-first century, something unprecedented began to happen. Machines, once simple tools designed to execute fixed instructions, started to demonstrate behaviors that resembled aspects of human cognition.

Computers began recognizing images, understanding speech, translating languages, and generating written text. Algorithms could identify patterns within enormous datasets and make predictions with astonishing accuracy.

What began as experimental research gradually evolved into a technological transformation.

Artificial intelligence had crossed an important threshold.

For the first time in the history of our planet, intelligence was no longer limited to biological organisms.

It had begun to emerge within machines.

A New Form of Intelligence

Artificial intelligence does not think the way humans do. Its processes are mathematical rather than biological, driven by neural networks, optimization algorithms, and enormous volumes of data.

Yet the outcomes can sometimes appear strikingly similar to human reasoning.

An AI system can analyze complex problems, generate solutions, and improve its performance through learning. It can synthesize knowledge across disciplines and perform calculations at speeds far beyond human capability.

In certain domains, machines have already surpassed human experts.

Artificial intelligence systems can defeat world champions in strategic games, analyze medical scans with extraordinary accuracy, and process financial information across global markets in real time.

These achievements represent more than incremental technological progress.

They represent the early stages of a new form of intelligence.

The Acceleration of Discovery

One of the defining characteristics of artificial intelligence is its potential to accelerate human discovery.

Throughout history, progress has often been limited by the pace of human reasoning and experimentation. Scientific breakthroughs required years of research and analysis.

Artificial intelligence has the potential to dramatically accelerate that process.

Machine learning systems can analyze enormous datasets, simulate complex environments, and identify patterns that might remain invisible to human researchers.

This capability could help scientists discover new medicines, develop advanced materials, and better understand the fundamental laws of nature.

The implications are profound.

If intelligence can be scaled through machines, the rate of discovery itself may accelerate.

Opportunity and Uncertainty

Every transformative technology introduces both opportunity and uncertainty.

The printing press reshaped communication and knowledge distribution. The steam engine powered the industrial revolution. The internet connected billions of people across the globe.

Artificial intelligence may represent a transformation of similar magnitude.

It could increase productivity, expand scientific understanding, and improve quality of life for billions of people.

At the same time, it raises important questions.

How should societies govern powerful intelligent systems? How will economies adapt as machines perform increasingly complex tasks? And what responsibilities do we carry as creators of a new form of intelligence?

These questions will define the coming decades.

Standing at the Beginning

Today, humanity stands at the beginning of a technological era that may fundamentally alter our relationship with machines.

AI is still evolving. Its capabilities continue to expand, and its long-term impact remains uncertain.

But one fact is already clear.

The age of machine intelligence has begun.

And the decisions made during the early years of this revolution will shape the future of civilization.

The chapters that follow explore the technologies, industries, and global forces driving this transformation, and what they may mean for the decades ahead.

The story of artificial intelligence is not just a story about machines.

It is a story about the future of human progress.

PART I

THE RISE OF MACHINE INTELLIGENCE

The Big Idea

For most of human history, intelligence belonged exclusively to biological life.

Every scientific discovery, technological breakthrough, and economic system was created through the reasoning ability of human minds. Tools amplified physical labor, but intelligence itself remained limited to the human brain.

AI is beginning to change that.

Advances in machine learning, computing infrastructure, and massive data systems have enabled machines to perform tasks that once required human reasoning. Algorithms can now analyze complex datasets, generate written content, recognize images, and assist with scientific research.

What makes this moment extraordinary is not simply that machines can perform useful tasks.

It is that intelligence itself is becoming a scalable technology.

Instead of being limited to individual human minds, intelligence can now be embedded within software systems capable of operating at global scale.

Understanding how this transformation began is essential.

The chapters in this section explore the origins of modern artificial intelligence, the technological breakthroughs that enabled recent advances, and the infrastructure required to support machine intelligence.

Together, these developments reveal how a field once confined to academic laboratories has rapidly become one of the most powerful technological forces shaping the modern world.

The rise of machine intelligence marks the beginning of a new technological era.

And like every major technological revolution before it, the consequences will extend far beyond the laboratories where the technology was first developed.

Why This Matters Now

Artificial intelligence did not emerge overnight.

The technologies shaping today's AI revolution are the result of decades of scientific research, technological experimentation, and breakthroughs in computing power. Yet in recent years, these developments have begun accelerating at an extraordinary pace.

For the first time at global scale, machines are capable of performing tasks that resemble human reasoning, language generation, and complex problem solving.

Understanding how this transformation began is essential.

The chapters in this section explore the origins of modern artificial intelligence, the global competition surrounding its development, and the massive infrastructure required to support machine intelligence.

Together, they reveal how a field once confined to academic laboratories has rapidly become one of the most important technological forces shaping the modern world.

"The question of whether machines can think is about as relevant as the question of whether submarines can swim." —Edsger Dijkstra

For most of human history, intelligence belonged to the human mind.

Every invention, every discovery, every civilization was built on the limits of human thought.

Then something extraordinary happened.

Machines began to learn.

They began to recognize patterns, analyze information, and solve problems once thought to require human reasoning.

What started as a technological experiment is rapidly becoming something far larger.

Intelligence itself is becoming a scalable resource.

And the consequences may reshape the global economy.

CORE THESIS

Artificial intelligence transforms intelligence itself into an economic resource.

For centuries, economic growth depended on labor, capital, and natural resources.

Today, a new factor is emerging: scalable intelligence.

As AI systems learn, reason, and solve problems, intelligence can be deployed across industries

in the same way machines once multiplied physical labor.

The result may be the rise of a new economic era:

The Intelligence Economy.

THE INTELLIGENCE ECONOMY FRAMEWORK

The intelligence revolution is built on several layers of technological and economic infrastructure.

Layer 1 — Compute Infrastructure

Advanced chips, cloud computing, and global data centers that power modern AI systems.

Layer 2 — Foundation Models

Large-scale AI models capable of reasoning, learning, and generating new information.

Layer 3 — Platforms

Technology platforms that allow developers and companies to build applications on top of AI systems.

Layer 4 — Applications

AI-powered software transforming industries such as medicine, finance, robotics, and education.

Layer 5 — Industry Transformation

The large-scale economic impact as intelligence becomes embedded across the global economy.

THE INTELLIGENCE ECONOMY STACK

INDUSTRY TRANSFORMATION

AI embedded across sectors

▲

APPLICATIONS

AI products & services

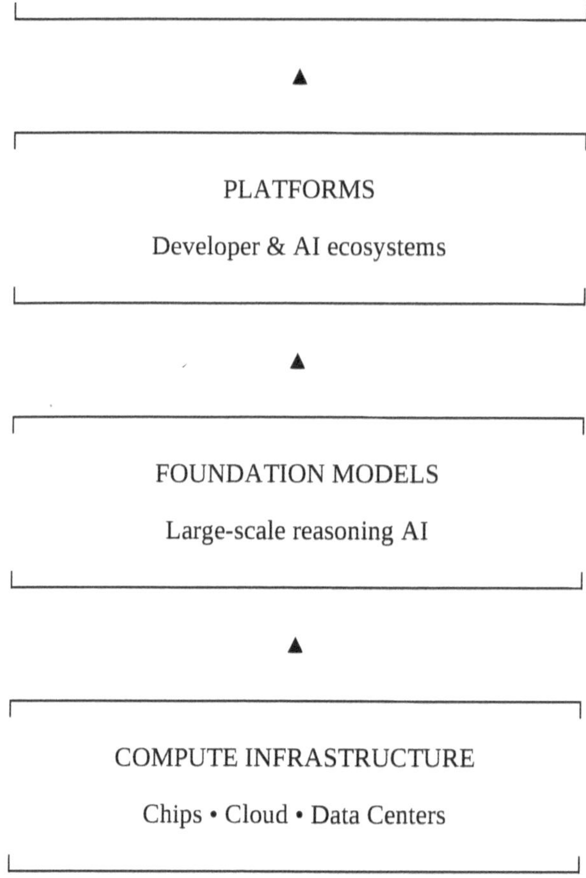

FIGURE: The layered architecture of the Intelligence Economy.

THE $100 TRILLION OPPORTUNITY

For most of human history, economic growth was constrained by human labor.

The Industrial Revolution multiplied physical labor with machines.

The Digital Revolution multiplied information with computers.

The Intelligence Revolution may multiply intelligence itself.

If intelligence becomes scalable through artificial intelligence systems, the productive capacity of the global economy could expand dramatically.

Some economists and technologists believe this shift could unlock tens of trillions of dollars in new economic value.

This possibility raises one of the most important questions of our time: What happens when intelligence becomes an economic resource?

Chapter 1

The Threshold

For most of human history, intelligence belonged exclusively to the human mind. That assumption is now being rewritten.

In the early decades of the twenty-first century, millions of people began interacting directly with systems capable of writing, reasoning, generating ideas, and explaining complex concepts in plain language. What had once required years of training could suddenly be accessed in seconds. Students used it to learn. Entrepreneurs used it to build. Professionals used it to accelerate work that had once required teams.

To many, this felt sudden. It was not. What appeared as a breakthrough was the visible crossing of a threshold decades in the making.

When History Changes

History does not change only when something new appears. It changes when a capability crosses from rarity into scale, when what was once limited becomes widely usable, economically valuable, and institutionally decisive. That is the threshold the world has now crossed.

For centuries, intelligence occupied a singular place in human civilization. It could be cultivated, trained, refined, and organized, but it could not be replicated outside the human mind. Societies built their institutions around that reality. Education formed judgment. Companies organized expertise. Governments relied on human interpretation to manage complexity. Markets rewarded the ability to think, decide, analyze, and coordinate under uncertainty. Machines extended physical force and widened the reach of communication. But cognition remained human.

That arrangement is now under pressure. The significance of this moment lies not in the appearance of new tools, but in the fact that intelligence has crossed into practical economic life at scale. What was once confined to research labs and experimental settings has entered public and commercial reality. The issue is not novelty. The issue is deployability. Intelligence is no longer functioning only as a human trait. It is becoming a scalable capability. And once a society crosses a threshold like this, it does not simply gain a new tool. It begins to reorganize around a new condition of power.

The End of a Human Monopoly

For most of recorded history, producing intelligence was slow, expensive, and inseparable from human development. To create a skilled mind required time, discipline, institutions, and sustained investment. Knowledge accumulated through effort and experience, and the supply of high-level cognitive labor remained constrained by the limits of human formation itself. This gave intelligence an economic character defined by scarcity.

Scarcity shaped wages, hierarchy, and institutional design. It determined who could interpret complexity, who could lead, and who could operate at scale. Entire professions existed because specialized knowledge was difficult to acquire and replicate. Entire industries depended on informational asymmetries that only trained individuals could navigate. Human intelligence was not merely valued. It was priced.

Artificial intelligence changes that equation by altering the relationship between cognition and scarcity. This does not make human intelligence irrelevant. It means that tasks involving language, synthesis, pattern recognition, analysis, and structured reasoning can now be performed or supported by machine systems at extraordinary scale. Once that becomes possible, intelligence shifts from being a human attribute to an operational input. Economies do not remain stable when a core input changes form. When energy changed form, industrial society emerged. When information changed form, the digital economy emerged. When intelligence changes form, the structure of work and value creation begins to change with it.

The Long Convergence

To many observers, this transformation appeared sudden. It was not. Major turning points are rarely as sudden as they seem. They are the result of multiple forces maturing at once, and artificial intelligence became consequential not because of one invention or one company, but because several currents developed over decades and converged.

The formal pursuit began in the mid-twentieth century, when early computer scientists asked whether machines could simulate human thinking. Alan Turing framed the challenge practically: if a machine could produce responses indistinguishable from a human, it could be considered intelligent. At the time, computers occupied entire rooms and possessed no understanding of language or reasoning. Yet the trajectory had begun.

Progress through the 1950s and 1960s produced systems that could solve problems and play structured games. Optimism ran high and funding flowed. But rule-based approaches could not handle real-world complexity. Human cognition involves context, ambiguity, and judgment, qualities that proved impossible to encode into fixed rules. By the late 1970s, expectations had collapsed. Funding declined, research slowed, and artificial intelligence entered what became known as the AI Winter.

What broke the stalemate was not a single breakthrough. It was the arrival of two critical ingredients the field had always lacked: data and computing power. The internet produced vast quantities of machine-readable information, allowing learning systems to train on real-world examples rather than programmer-written rules. At the same time, graphics processing units originally designed for video games proved ideally suited for training neural networks, capable of performing thousands of calculations simultaneously. Combined, these forces created the conditions for a genuine leap.

That leap arrived in 2012, when a deep learning system demonstrated dramatic improvements in image recognition at a global competition called ImageNet. The results shocked the research community. Within years, deep learning was outperforming traditional software across speech recognition, language translation, medical imaging, and autonomous driving. Artificial intelligence had become commercially valuable.

The most consequential development was still ahead. Around 2017, researchers introduced the transformer, an architecture that could analyze relationships across entire sequences of language simultaneously, rather than processing text step by step. This made possible large language models trained on vast internet datasets, capable of writing, reasoning, translating, and generating code all within a single architecture. For the first time, machines were demonstrating something resembling flexible, general-purpose intelligence.

Computation, data, modeling, infrastructure, and capital had all converged. What had been fragmented became integrated. What had been theoretical became practical. What had been experimental became scalable. This is how thresholds are crossed, not through a single event, but through accumulation. The visible breakthrough is not the beginning. It is the confirmation.

Why Society Noticed Late

Human beings rarely experience transformation at its structural level. They encounter it through surface effects, new tools, new behaviors, new efficiencies, new anxieties, but not the underlying logic. This delay between reality and recognition is common in every age of transition.

Industrialization was first experienced through changing labor and urban life, not as economic reorganization. The internet was first experienced as websites and communication, not as civilizational infrastructure. Artificial intelligence is being experienced the same way, as a tool, a convenience, a novelty. But its significance lies deeper.

Society noticed when machine cognition became visible enough to enter daily conversation. The transformation had already begun once machine systems became capable of economically useful cognitive work at scale. Public awareness followed operational reality. If artificial intelligence is understood merely as a striking new tool, its significance may seem temporary. If it is understood as the point at which intelligence became deployable as infrastructure, its significance becomes structural. Applications will change. Companies will rise and fall. Interfaces will evolve. But the condition remains. Once crossed, the threshold cannot be reversed by changes in branding, leadership, or fashion.

Intelligence as Infrastructure

The most important fact about this era is not that machines have become more impressive. It is that intelligence is becoming infrastructure.

Infrastructure transforms societies by embedding itself into systems, reducing costs, expanding scale, and reshaping expectations. Roads transformed movement. Electricity transformed industry. Networks transformed information. Artificial intelligence is beginning to transform cognition. When intelligence becomes infrastructural, it becomes distributable, repeatable, and embedded across organizations. This does not eliminate human relevance. But it changes how human ability is organized and valued.

As intelligence becomes more abundant, scarcity does not disappear. It moves — toward judgment, trust, originality, coordination, and strategic control. It moves toward ownership of data, infrastructure, and the ability to direct systems rather than merely participate in them. Economies reorganize whenever abundance replaces scarcity. This is not only a technology story. It is a reallocation of value.

The Pressure on Labor, Institutions, and Power

The consequences of this shift will not unfold evenly. Some sectors will see augmentation before disruption. Others will undergo reorganization before displacement. In many cases, the first effect will not be elimination but compression of roles, margins, and timelines. Tasks once requiring teams may require fewer people. Work once protected by informational scarcity faces new competition.

Education must respond because institutions designed to prepare people for scarce expertise now operate in a world where some expertise can be simulated or commoditized. Firms must respond because competitive advantage will depend less on access to routine cognition and more on control over systems and execution. Governments must respond because intelligence capacity increasingly shapes national strength and geopolitical positioning. Individuals must respond because the terms of cognitive competition are changing.

Every age of transformation produces both overstatement and denial. Both misunderstand transition. Structural change is rarely instantaneous, but it is no less real for unfolding unevenly. A society built on the scarcity of human cognition cannot remain unchanged when cognition becomes scalable. The rate may vary. The political response may vary. The winners and losers may vary. But the pressure is already present.

What This Book Is About

This book is not an argument for hype, nor for fear. It is an effort to understand the deeper meaning of a transformation already underway.

The central claim is simple: when intelligence becomes scalable, the systems built around its scarcity begin to change. That change reaches across labor markets, economic organization, national power, and human identity. It alters how value is created, who captures it, and what institutions remain durable. The chapters that follow examine these consequences not as isolated disruptions, but as expressions of a deeper restructuring.

The world has not merely encountered a new tool. It has crossed a threshold. Thresholds matter because they alter what becomes normal. Old assumptions lose stability. Systems designed for one condition must adapt to another. What once seemed fixed begins to move. The future does not arrive all at once, but it does arrive through these crossings.

And once crossed, they reorder everything that follows.

Key Takeaways

- Intelligence has crossed a critical threshold, from scarce human capability to scalable system capability.
- The significance of artificial intelligence is not novelty, but deployability at scale within economic systems.
- Societies organize themselves around the conditions of scarcity; when those conditions change, institutions must reorganize.
- Artificial intelligence does not eliminate human relevance, but it redefines how human intelligence is valued and applied.

- Every major transformation appears sudden, but is actually the result of long-term convergence across multiple forces.
- Public awareness lags behind structural change; by the time society notices, the transformation is already underway.
- As intelligence becomes abundant, scarcity shifts toward judgment, control, coordination, and system ownership.
- This is not merely a technological shift; it is a reallocation of economic value and power.
- The transition will not be uniform; it will manifest as compression, restructuring, and uneven disruption across sectors.
- Once a threshold of this magnitude is crossed, it cannot be reversed, only adapted to.

Expert Insight

The defining mistake in interpreting artificial intelligence is treating it as a tool rather than a condition.

Tools can be adopted or ignored.

Conditions reshape behavior regardless of preference.

Artificial intelligence represents a shift in the underlying economics of cognition. When a capability that was once scarce becomes abundant, the systems built around its scarcity, education, labor markets, corporate structure, and even geopolitical power lose stability.

This is why the comparison to past technologies is often insufficient. Machines once replaced physical labor. Software once accelerated information processing. Artificial intelligence is different. It directly engages with cognitive work, the very layer upon which modern economies and institutions are built.

The implication is not immediate displacement, but structural pressure.

Organizations that understand this shift early will reorganize around leverage. Those that do not will continue operating under assumptions that no longer hold. The same is true for individuals, industries, and nations.

The threshold has already been crossed.

The only remaining variable is who adapts first and who absorbs the cost of delay.

Strategic Question

If intelligence is no longer scarce, then:

Where does value concentrate next and what position do you occupy in that new structure?

More specifically:

- Are you competing on tasks that can be replicated, or on judgment that cannot?
- Are you participating in systems, or controlling the systems themselves?
- Are you adapting your model of work, or defending one built on outdated assumptions?

This is no longer a question of awareness.

It is a question of positioning.

Looking Ahead

The chapters that follow move beyond recognition and into consequence.

If Chapter 1 establishes that a threshold has been crossed, the rest of the book examines what that crossing produces.

- How labor markets reorganize when cognitive work becomes scalable
- How economic value shifts when intelligence becomes embedded in systems
- How power concentrates around those who control infrastructure rather than those who operate within it

- How institutions, education, corporations, and governments respond under pressure
- How individuals navigate a landscape where traditional advantages erode and new ones emerge

The central question is no longer whether artificial intelligence will matter.

It is how deeply it will restructure the systems we depend on and how quickly.

The threshold is not the end of the story.

It is the beginning of everything that follows.

Chapter 2

The Global AI Arms Race

Nations once competed for land, oil, and military strength. Today, the most important race on Earth may be the race to build the most powerful artificial intelligence.

A New Technological Competition

AI is no longer simply a scientific pursuit.

It has become a geopolitical competition.

Nations across the world now recognize that AI may become one of the most strategically important technologies of the 21st century. Much like nuclear energy in the mid-twentieth century or the internet in the 1990s, artificial intelligence carries the potential to reshape economic power, military capability, and global influence.

Governments, corporations, and research institutions are investing enormous resources to ensure they do not fall behind.

This has triggered what many analysts now describe as the global AI arms race.

Unlike earlier technological rivalries, however, this competition is not limited to military capability. Artificial intelligence has implications across nearly every sector of society, from healthcare and finance to education, manufacturing, and transportation.

The stakes are therefore much broader.

The country that leads in AI development may gain advantages not only in defense but also in economic growth, scientific discovery, and technological innovation.

The New Strategic Resource: Intelligence

Throughout history, power has often been determined by control over critical resources.

In ancient civilizations, the key resource was land and agriculture. Nations capable of producing large quantities of food could support larger populations and armies.

During the industrial revolution, the strategic resources shifted to coal, steel, and oil. Countries that controlled these materials were able to build factories, railroads, and military infrastructure.

In the late twentieth century, information networks and computing technology became the new engines of global power. Nations that developed advanced digital infrastructure dominated the emerging technology economy.

Today, the emerging strategic resource is machine intelligence.

Artificial intelligence systems have the potential to influence nearly every major sector of society.

Economic productivity

Scientific research

Military operations

Cybersecurity

Healthcare innovation

Transportation systems

Energy infrastructure

The countries that lead in AI development may gain significant advantages in both economic growth and national security.

For this reason, governments are paying close attention.

The United States and the AI Ecosystem

The United States currently holds one of the strongest positions in artificial intelligence development.

This advantage stems from a unique ecosystem that combines research universities, technology companies, venture capital, and advanced computing infrastructure.

American universities have long been leaders in computer science and artificial intelligence research. Institutions such as Stanford, MIT, Carnegie Mellon, and Berkeley have produced many of the researchers responsible for major AI breakthroughs.

Equally important are the technology companies that have emerged from this ecosystem.

Several of the world's most influential AI organizations operate within the United States, including companies developing large-scale machine learning systems and cloud computing platforms.

These companies invest billions of dollars annually into artificial intelligence research and development.

They also control much of the computational infrastructure required to train advanced AI systems.

But the strength of the American AI ecosystem extends beyond large corporations.

Startups, venture capital firms, and open-source communities also play a major role in accelerating innovation.

This decentralized innovation environment allows new ideas to emerge rapidly and spread across the technology industry.

The Hardware Behind AI Leadership

Artificial intelligence requires more than just algorithms.

It requires enormous computational power.

Training modern AI models can involve billions or even trillions of mathematical operations performed across vast neural networks.

To support this level of computation, technology companies rely on specialized processors designed specifically for machine learning workloads.

The trajectory of specialized hardware illustrates how adjacent industries can become foundational to an intelligence revolution. One early example of this pattern was NVIDIA a company that began primarily in graphics computing but whose processors proved uniquely suited to the mathematical operations that machine learning required.

Originally designed for rendering video game graphics, its processors could perform thousands of parallel calculations simultaneously exactly the kind of workload that training neural networks demanded.

Over time, its GPUs became the backbone of AI infrastructure around the world.

Large AI models are often trained using thousands of GPUs operating together in massive computing clusters.

These clusters represent billions of dollars in hardware investment.

As artificial intelligence continues to expand, the demand for advanced AI chips has grown dramatically.

In many ways, control over advanced semiconductors has become a critical component of the global AI competition.

China's AI Ambitions

While the United States currently leads in many aspects of AI development, it is not the only nation investing heavily in the technology.

China has also made artificial intelligence a national priority.

Over the past decade, Chinese technology companies and government agencies have invested billions of dollars into AI research and development.

Major Chinese technology firms have built extensive AI research divisions focused on areas such as:

Computer vision

Natural language processing

Autonomous vehicles

Facial recognition technology

Smart city infrastructure

In 2017, the Chinese government released a national strategic plan outlining its ambition to become the global leader in artificial intelligence by 2030.

The plan called for massive investments in research laboratories, AI startups, and advanced computing infrastructure.

China also benefits from several structural advantages in AI development.

The country has a large population generating enormous quantities of data, particularly in areas such as mobile commerce and digital payments.

It also possesses a rapidly growing technology workforce.

Together, these factors have positioned China as one of the most significant competitors in the global AI landscape.

Europe and the Regulatory Approach

While much of the technological competition in AI is often framed as a rivalry between the United States and China, Europe has taken a somewhat different approach.

European nations have also invested in artificial intelligence research, particularly through collaborations between universities and public research institutions.

However, Europe has placed a strong emphasis on regulation and ethical oversight of AI technologies.

European policymakers have expressed concerns about issues such as:

Algorithmic bias

Privacy protection

Transparency in automated decision-making

The societal impact of automation

These concerns have led to the development of regulatory frameworks designed to ensure that AI technologies are developed responsibly.

While some critics argue that regulation could slow innovation, others believe it may help build public trust in AI systems.

In the long term, Europe's approach could shape global standards for responsible AI development.

The Infrastructure War

Behind the scenes, the most intense competition in artificial intelligence may not involve algorithms at all.

It involves infrastructure.

Training advanced AI models requires massive computing clusters composed of thousands, sometimes tens of thousands, of specialized processors operating together.

These clusters consume enormous amounts of electricity and require sophisticated cooling systems to operate efficiently.

Technology companies are now building some of the largest computing facilities ever constructed.

Major cloud providers are expanding global networks of AI data centers designed specifically for training and operating large-scale machine learning systems.

These facilities represent billions of dollars in infrastructure investment.

But they also represent something more important.

Control over the computational engines of intelligence.

Whoever controls the infrastructure capable of training advanced AI models may ultimately control the pace of future AI development.

The Talent Competition

Another critical factor in the AI race is talent.

Artificial intelligence research requires highly skilled scientists, engineers, and mathematicians capable of developing complex machine learning systems.

Countries and corporations are therefore competing aggressively to attract top AI researchers.

Universities produce many of these experts, but the demand for their expertise has grown dramatically.

Technology companies frequently recruit leading researchers directly from academia, offering significant resources and research funding.

This movement of talent between universities and industry has accelerated the pace of AI innovation.

At the same time, it has intensified the global competition for expertise.

Why AI Could Reshape the Global Economy

The economic implications of artificial intelligence may be enormous.

AI systems have the potential to dramatically increase productivity across many sectors by automating complex analytical tasks and accelerating research and development.

Industries that could be significantly transformed include:

Healthcare

Finance

Manufacturing

Transportation

Energy

Education

Media and entertainment

Some economists estimate that AI could contribute trillions of dollars in additional economic output over the coming decades.

At the same time, AI may create entirely new industries that do not yet exist.

Just as the internet gave rise to companies such as Google, Amazon, and social media platforms, the AI revolution may produce new categories of businesses built around machine intelligence.

For entrepreneurs, this represents both an opportunity and a challenge.

The pace of technological change is accelerating, and the companies that succeed will likely be those that learn to integrate artificial intelligence into their products and services.

A Technological Turning Point

The global race for artificial intelligence leadership is still in its early stages.

Breakthroughs continue to emerge in areas such as:

Autonomous robotics

Advanced reasoning models

AI-driven scientific discovery

Machine-assisted engineering

Human-AI collaboration systems

No one can predict exactly how these developments will unfold. But one thing is becoming increasingly clear. AI is not just another technological trend.

It is rapidly becoming a foundational technology that could shape the balance of economic and geopolitical power in the decades ahead.

And the race to build that future has already begun.

Key Takeaways

• Artificial intelligence has become a central focus of global technological competition.

• Nations view AI leadership as critical for economic growth, military capability, and technological influence.

• The United States, China, and Europe represent major centers of AI development with different strategic approaches.

• Semiconductor technology and computing infrastructure are essential resources in the AI race.

• Competition for AI talent and research expertise is intensifying globally.

Expert Insight

Throughout history, technological revolutions have often reshaped global power structures. The development of nuclear technology influenced geopolitical dynamics during the twentieth century. Similarly, the internet revolution transformed communication, commerce, and information exchange across the globe.

Artificial intelligence may represent a comparable shift.

Nations capable of developing advanced AI infrastructure, research ecosystems, and skilled workforces may gain significant economic advantages. These advantages extend beyond traditional industries, influencing fields such as cybersecurity, defense systems, biotechnology, and advanced manufacturing.

Because of this broad impact, the global race for AI leadership is likely to remain one of the defining technological competitions of the coming decades.

Strategic Question

If artificial intelligence becomes the most important strategic technology of the twenty-first century, which nations will lead its development, and how will that leadership reshape global power?

Will the future balance of influence be determined by military strength, economic resources, or by control over intelligent technologies?

Looking Ahead

While algorithms and research breakthroughs often dominate public discussion about AI, the true engines of artificial intelligence lie beneath the surface.

The next chapter explores the massive physical infrastructure required to power modern AI systems: data centers, specialized processors, and global computing networks.

Chapter 3

The Infrastructure Behind Intelligence

Every technological revolution is built on invisible foundations. The intelligence revolution rests on data centers, chips, and vast oceans of data.

The Hidden Foundation of Artificial Intelligence

When people think about artificial intelligence, they often imagine software.

They picture algorithms analyzing data, chat systems generating language, or machine-learning models recognizing images.

But behind every advanced AI system lies something far less visible and far more expensive.

Infrastructure.

Artificial intelligence does not exist in isolation. It runs on enormous networks of computers, specialized processors, and global data centers. These systems form the physical foundation that makes modern AI possible.

Without this infrastructure, even the most advanced algorithms would remain theoretical concepts.

Every time a user asks an AI system a question, requests a generated image, or analyzes data with machine learning tools, an enormous amount of computing power operates behind the scenes.

Thousands of processors perform calculations simultaneously, drawing upon vast networks of servers distributed across global data centers.

Artificial intelligence may feel intangible, but its foundation is deeply physical.

The AI revolution is built on machines.

The Computational Hunger of AI

Training modern artificial intelligence systems requires staggering amounts of computation.

Large neural networks often contain billions, sometimes hundreds of billions, of parameters. These parameters represent the internal structure that allows the system to learn patterns within data.

During training, the model repeatedly adjusts these parameters through mathematical operations performed across enormous datasets.

For example, a large language model may process trillions of words during training. Each pass through the dataset requires the system to perform billions of calculations.

The total number of computations required can reach into the quintillions.

Performing these calculations using traditional processors would take an impractical amount of time.

To solve this problem, researchers began using specialized processors capable of performing large numbers of mathematical operations simultaneously.

These processors are known as graphics processing units, or GPUs.

The Rise of GPU Computing

Graphics processing units were originally designed for an entirely different purpose.

During the early days of computer gaming, developers needed hardware capable of rendering complex visual scenes in real time. GPUs were created to perform the massive number of calculations required to generate these graphics.

What researchers later discovered was that the same type of computation required for graphics rendering also worked extremely well for machine learning.

Neural networks rely heavily on matrix multiplication, a mathematical operation that GPUs can perform extremely efficiently.

Unlike traditional CPUs, which execute a small number of complex tasks sequentially, GPUs can perform thousands of simpler operations simultaneously.

This parallel processing capability makes them ideal for training neural networks.

As machine learning research accelerated, GPUs gradually became the dominant hardware platform for AI development.

Technology companies began building enormous computing clusters composed of thousands of GPUs connected together through high-speed networks.

These clusters became the engines powering modern artificial intelligence.

The Emergence of AI Supercomputers

As AI models grew larger, the scale of the required infrastructure increased dramatically.

Training cutting-edge AI systems now often requires what can best be described as AI supercomputers.

These systems consist of thousands, sometimes tens of thousands, of GPUs operating together as a single computational unit.

They require specialized networking hardware capable of transmitting massive amounts of data between processors at extremely high speeds.

They also require advanced cooling systems to prevent the hardware from overheating during intensive training operations.

The cost of building these systems can reach hundreds of millions of dollars.

Only a small number of organizations currently possess the resources required to train the largest AI models.

These organizations include major technology companies, government research laboratories, and a handful of well-funded startups.

In many ways, access to large-scale computing infrastructure has become one of the defining factors in the AI race.

The Global Data Center Network

The infrastructure required for artificial intelligence extends far beyond individual computing clusters.

Behind modern AI systems lies a vast global network of data centers.

These facilities contain rows of server racks packed with processors, storage systems, and networking equipment.

Inside a large data center, tens of thousands of servers may operate simultaneously, processing enormous volumes of data.

These facilities require vast amounts of electricity.

In some cases, a single data center can consume as much power as a small city.

Maintaining these systems also requires advanced cooling technologies to dissipate the heat generated by thousands of high-performance processors operating continuously.

Technology companies have built global networks of these facilities in order to provide cloud computing services.

These networks allow organizations around the world to access powerful computing resources without building their own infrastructure.

Cloud computing has therefore become one of the key drivers of AI innovation.

Cloud Computing and the Democratization of AI

One of the most important developments in the AI ecosystem has been the rise of cloud computing platforms.

These platforms allow businesses, researchers, and startups to rent access to powerful computing infrastructure through the internet.

Instead of purchasing expensive hardware, organizations can run machine learning workloads on remote servers hosted by cloud providers.

This model has dramatically lowered the barrier to entry for artificial intelligence development.

A small startup team can now train sophisticated machine learning models using infrastructure that would have been impossible for them to build independently.

As a result, AI innovation is no longer limited to large corporations.

Researchers, entrepreneurs, and independent developers can all experiment with machine learning tools.

This democratization of AI infrastructure has accelerated the pace of innovation across the industry.

The Semiconductor Supply Chain

While cloud computing platforms make AI infrastructure accessible, they still depend on a complex global supply chain.

At the center of this supply chain lies the semiconductor industry.

Modern AI processors require extremely advanced manufacturing techniques.

The transistors inside these chips are measured in nanometers, or billionths of a meter.

Producing these chips requires some of the most sophisticated manufacturing equipment ever created.

Only a small number of companies possess the technological capabilities required to produce cutting-edge semiconductor devices.

As artificial intelligence becomes increasingly important, semiconductor manufacturing has emerged as a strategic industry.

Governments around the world are investing heavily in domestic chip production in order to secure access to these critical technologies.

The semiconductor supply chain has therefore become one of the most important battlegrounds in the global AI race.

The Energy Challenge

The rapid expansion of artificial intelligence infrastructure has also introduced a new challenge.

Energy consumption.

Training and operating large AI systems requires enormous amounts of electricity.

As AI adoption continues to grow, the energy demands of data centers are expected to increase significantly.

Some analysts believe that AI computing could eventually account for a substantial share of global electricity consumption.

This has prompted researchers and engineers to explore new ways of improving energy efficiency in AI systems.

Potential solutions include:

More efficient semiconductor designs

Advanced cooling technologies

Specialized AI processors

Renewable energy-powered data centers

Some companies are even exploring the possibility of building data centers near large sources of renewable energy in order to reduce environmental impact.

Managing the energy demands of AI infrastructure may become one of the major engineering challenges of the coming decades.

The Economics of AI Infrastructure

Building large-scale AI infrastructure requires enormous financial investment.

The cost of constructing advanced data centers, developing specialized processors, and operating massive computing clusters can reach billions of dollars.

For technology companies, however, these investments are often justified by the potential economic impact of artificial intelligence.

AI-powered services can generate revenue across many sectors, including:

Cloud computing platforms

Software development tools

Data analytics services

Autonomous systems

Enterprise AI applications

As demand for these services grows, companies are racing to expand their infrastructure capacity.

The organizations capable of building the most powerful computing networks may gain a significant competitive advantage in the AI economy.

The Foundations of the AI Era

The artificial intelligence revolution is often described in terms of algorithms and software.

But the reality is far more complex.

Behind every AI breakthrough lies an enormous physical infrastructure composed of processors, networks, data centers, and global supply chains.

This infrastructure represents the hidden foundation of machine intelligence.

Just as the industrial revolution required railroads, factories, and power plants, the AI revolution requires computing clusters, semiconductor manufacturing, and energy networks.

Understanding this infrastructure is essential to understanding the future of artificial intelligence.

Because the race to build intelligent machines is also a race to build the systems that power them.

And that race is only beginning.

Key Takeaways

• Artificial intelligence relies on enormous computing infrastructure.

• GPU processors and specialized AI chips power modern machine learning systems.

• Data centers and cloud computing networks form the backbone of global AI operations.

• Semiconductor manufacturing has become a strategic industry in the AI economy.

• Energy consumption and infrastructure scalability present major challenges for future AI growth.

Expert Insight

Behind every AI breakthrough lies an extensive physical system composed of hardware, energy networks, and global supply chains. While software innovations often capture public attention, the infrastructure required to train and operate large AI models represents one of the most significant engineering achievements of the digital age.

Building and maintaining this infrastructure requires massive investment. Technology companies are constructing enormous data centers capable of supporting thousands of high-performance processors operating simultaneously.

As artificial intelligence continues to evolve, access to large-scale computing infrastructure may become one of the most important competitive advantages in the technology sector.

Strategic Question

If artificial intelligence depends on massive computing infrastructure and specialized hardware, who should control the systems that power machine intelligence?

Will access to advanced AI infrastructure remain concentrated within a small number of corporations and governments, or will it become broadly accessible to innovators around the world?

Looking Ahead

With infrastructure expanding rapidly, the capabilities of AI systems are also evolving.

The next chapter explores a new development in artificial intelligence: AI agents, systems capable of performing complex tasks autonomously rather than simply responding to prompts.

PART II

THE EXPANSION OF ARTIFICIAL INTELLIGENCE

The Big Idea

The early decades of artificial intelligence were defined primarily by research.

Scientists developed algorithms capable of recognizing patterns, analyzing data, and solving narrow technical problems. For many years these systems remained largely confined to laboratories and specialized technology companies.

That situation is now changing.

AI is moving beyond experimental research and entering everyday life.

Businesses are integrating machine learning into their operations. Entrepreneurs are building new companies around intelligent software systems. Professionals across many industries are beginning to use AI tools to accelerate research, automate routine tasks, and enhance decision-making.

This shift marks an important turning point.

AI is no longer simply a field of computer science.

It is becoming a general-purpose technology capable of transforming how work is performed, how businesses operate, and how innovation occurs.

One of the most important developments in this transition is the rise of AI agents, systems capable of performing complex tasks autonomously rather than simply responding to individual prompts. These systems can coordinate workflows, gather information, interact with digital tools, and generate outputs with minimal human direction.

As these capabilities expand, the relationship between humans and technology may begin to change.

Instead of operating software directly, people may increasingly delegate tasks to intelligent systems capable of acting on their behalf.

This transformation has profound implications for the workforce.

Artificial intelligence has the potential to automate certain types of cognitive labor while simultaneously increasing the productivity of individuals and organizations. Entire professions may evolve as humans learn to collaborate with intelligent machines.

The chapters in this section explore how artificial intelligence is expanding beyond research laboratories and becoming embedded in real-world systems.

They examine the emergence of AI agents, the transformation of work, and the early stages of what may become one of the largest economic expansions in modern history.

The AI revolution is no longer theoretical.

It has begun reshaping the world of work and innovation.

And its expansion is only just beginning.

Why This Matters Now

As artificial intelligence technologies matured, they began moving beyond research laboratories and into real-world applications.

Businesses, entrepreneurs, and researchers started experimenting with ways to apply machine learning to practical challenges across industries.

This phase marked a critical turning point.

Artificial intelligence was no longer just an experimental technology; it was becoming a tool capable of transforming how work is performed, how businesses operate, and how innovation unfolds.

The chapters in this section examine the rapid expansion of AI systems into everyday workflows, the rise of autonomous digital agents, and the growing ecosystem of startups building businesses around machine intelligence.

"AI is likely to be either the best or worst thing to happen to humanity."

Stephen Hawking

Chapter 4

The Rise of Artificial Intelligence Agents

Software used to follow instructions. Now it can reason, plan, and act on its own.

From Tools to Autonomous Systems

For most of the history of computing, software has functioned as a tool.

A user would provide instructions, and the software would execute those instructions. Whether calculating numbers in a spreadsheet, editing text in a document, or performing complex engineering simulations, traditional software systems relied on human direction.

Artificial intelligence began by enhancing those tools.

Early AI systems could recognize speech, translate languages, or classify images. These capabilities made software more powerful, but the basic structure of human-computer interaction remained the same.

Humans gave commands.

Machines produced outputs.

But a new category of AI systems is beginning to change that relationship.

These systems are known as AI agents.

Unlike traditional software tools, AI agents are designed to perform tasks autonomously. Instead of simply responding to a single request, they can plan, execute multi-step processes, and adapt their behavior based on new information.

In other words, they can act.

This shift from passive tools to active digital systems may represent one of the most important developments in the future of artificial intelligence.

What Is an AI Agent?

An AI agent is a system capable of taking actions in pursuit of a goal.

Rather than simply generating text or answering questions, an agent can perform sequences of tasks, interact with software tools, retrieve information, and make decisions along the way.

For example, imagine giving an AI system the following instruction:

"Research potential markets for a new technology startup and produce a report on the three most promising industries."

A traditional AI chatbot might generate a brief summary based on its training data.

An AI agent, however, could approach the task differently. It might:

Search online databases for market research reports

Analyze financial data for different industries

Compile statistics about growth trends

Generate a structured report summarizing its findings

Throughout this process, the agent would continuously evaluate the results of its actions and determine the next step required to complete the objective.

In effect, it would behave like a digital research assistant.

The Architecture of Agentic Systems

AI agents typically combine several technological components.

At their core is a reasoning model capable of understanding instructions and generating responses. These models are often large language models trained on vast datasets of text and code.

But the reasoning model alone is not enough.

For an AI system to function as an agent, it must also interact with external tools and data sources.

This may include:

Web browsers for retrieving information

Databases for storing knowledge

Programming interfaces that allow the AI to run software tools

Memory systems that allow the agent to track progress across tasks

By combining reasoning with tool use, AI agents can perform complex workflows.

They can gather information, analyze results, generate outputs, and revise their approach when necessary.

This ability to coordinate multiple actions is what distinguishes agents from simpler AI systems.

The Automation of Knowledge Work

The emergence of AI agents has significant implications for the nature of work.

Many modern professions involve tasks that follow predictable processes.

Researchers gather information from multiple sources. Analysts process financial data. Project managers coordinate workflows across teams. Customer support representatives respond to common questions.

These tasks require cognitive effort, but they often follow structured procedures.

AI agents are particularly well suited for this type of work.

Because they can perform sequences of actions and access multiple software systems, they may eventually automate many forms of knowledge work.

For example, an AI agent might:

Analyze customer feedback data and generate product improvement recommendations

Monitor financial markets and produce daily investment summaries

Conduct research for journalists or academic researchers

Assist lawyers by reviewing large collections of legal documents

In many cases, these systems will not completely replace human professionals.

Instead, they may function as assistants capable of performing large portions of routine work.

A single individual working alongside AI agents could potentially accomplish tasks that previously required entire teams.

AI Agents and Software Development

One of the earliest areas where AI agents have begun to show promise is software development.

Writing software often involves many interconnected tasks.

Developers must analyze requirements, design system architecture, write code, test functionality, identify bugs, and update documentation.

AI-powered coding assistants have already demonstrated the ability to help programmers generate code more quickly.

The next step is creating AI agents capable of handling larger portions of the development process.

Such agents might:

Generate initial software prototypes

Run automated tests

Identify and fix errors in code

Update documentation when changes occur

While human oversight will remain essential, AI agents could significantly accelerate the pace of software development.

This capability could lower the barrier to entry for new technology companies and increase the speed at which new digital products are created.

The Entrepreneurial Impact

The rise of AI agents may have a particularly strong impact on entrepreneurship.

Launching a new business typically requires a wide range of skills.

Entrepreneurs must conduct market research, develop products, create marketing materials, manage customer support, and analyze financial performance.

Traditionally, these tasks required building a team.

But AI agents may allow small groups, or even individuals, to perform many of these functions themselves.

An entrepreneur might use AI systems to:

Analyze potential market opportunities

Design product prototypes

Generate marketing campaigns

Automate customer interactions

This could dramatically reduce the cost and complexity of starting new companies.

The result may be a surge in innovation as more individuals gain the ability to launch new ventures.

Challenges and Limitations

Despite their promise, AI agents are still in the early stages of development.

Current systems can sometimes make errors when executing complex tasks. They may misinterpret instructions, rely on outdated information, or generate incorrect conclusions.

Ensuring reliability is therefore a major area of ongoing research.

Another challenge involves safety.

Autonomous systems capable of taking actions must be carefully designed to avoid unintended consequences.

Developers must ensure that agents follow appropriate rules, respect privacy protections, and operate within clearly defined boundaries.

Addressing these challenges will require improvements in both technology and governance.

But progress in this area is moving rapidly.

The Beginning of Autonomous Software

For decades, software has served as a tool that humans use to accomplish tasks.

AI agents represent the first step toward something different.

They are systems capable of operating with a degree of autonomy, executing complex workflows and interacting with the digital world on behalf of human users.

This shift may transform the relationship between people and technology.

Instead of manually operating software, users may increasingly delegate tasks to intelligent agents that can perform them independently.

As these systems become more capable, they may begin to function as digital collaborators, assisting with research, development, decision-making, and creative work.

In the long run, the rise of AI agents may fundamentally reshape the nature of digital work.

And like many aspects of the AI revolution, this transformation is only beginning.

Key Takeaways

• AI agents represent a shift from passive software tools to autonomous digital systems.

• These systems can plan, execute tasks, and interact with external tools and data sources.

• AI agents may automate many forms of structured knowledge work.

• Software development and research workflows are among the first areas influenced by agent systems.

• Autonomous software may significantly increase productivity across many industries.

Expert Insight

The emergence of AI agents signals an important shift in how humans interact with technology. Traditional software requires constant human direction, but agent systems can operate more independently.

This capability allows AI systems to coordinate complex workflows involving multiple steps and data sources. For example, an AI agent could gather information from databases, analyze the results, generate reports, and refine its conclusions based on feedback.

Although current agent systems remain experimental, they illustrate a future in which digital tools function less like instruments and more like collaborators.

Strategic Question

If AI agents can autonomously perform complex digital tasks, how will this change the way individuals interact with technology?

Will people continue to operate software directly, or will intelligent systems increasingly act on our behalf, researching, planning, and executing work across digital environments?

Looking Ahead

As AI systems become capable of performing increasingly complex tasks, their influence on the workforce will expand.

The next chapter explores how artificial intelligence may transform the nature of work itself.

Chapter 5

The Transformation of Work

Work has always evolved alongside technology, but machines that can think may reshape the very meaning of work itself.

A New Kind of Industrial Revolution

Throughout history, technological revolutions have reshaped the nature of work.

The agricultural revolution transformed human societies from nomadic hunter-gatherers into settled farming communities. The industrial revolution replaced manual craftsmanship with mechanized production. The digital revolution introduced computers and the internet, changing how information is created, stored, and shared.

Artificial intelligence may represent the next major transformation.

But unlike earlier technologies, AI has the potential to automate not only physical labor, but also certain forms of cognitive labor, specifically tasks that involve reasoning, analysis, and communication.

For centuries, machines have primarily replaced physical effort.

Factories replaced manual manufacturing. Agricultural machinery increased farming productivity. Construction equipment allowed large infrastructure projects to be completed more efficiently.

Human intellectual work, however, remained largely untouched.

Writing, planning, analysis, and decision-making were considered uniquely human abilities.

AI is beginning to challenge that assumption.

The Automation of Cognitive Tasks

Modern AI systems are capable of performing a variety of tasks that were once believed to require human intelligence.

These tasks include:

Writing reports and articles

Analyzing large datasets

Generating computer code

Summarizing documents

Providing customer support

Translating languages

Many of these functions are not complete replacements for human expertise, but they can dramatically accelerate the speed at which work is performed.

A financial analyst who once spent hours reviewing spreadsheets may now use AI systems to process data in seconds.

A marketing team that once required weeks to develop advertising copy may generate campaign concepts within minutes.

Software developers can now use AI-assisted tools to write portions of code automatically, allowing them to focus on system design and architecture.

These changes do not eliminate the need for human workers, but they do alter the structure of many professions.

The role of humans increasingly shifts from performing routine tasks to supervising, refining, and guiding AI systems.

Real-World Examples of AI in the Workplace

The transformation of work driven by artificial intelligence is already visible across multiple industries.

Software development provides one of the clearest examples. AI-assisted coding tools can now generate entire blocks of functional software based on simple instructions written in natural language. Developers who once spent hours writing routine code can now focus more on architecture, system design, and problem-solving.

Customer service is also undergoing rapid change. Many companies now deploy AI-powered systems capable of answering common questions, resolving basic support issues, and guiding customers through troubleshooting steps. These systems operate continuously and can assist thousands of users simultaneously.

In finance, machine learning models analyze massive volumes of financial data to identify unusual activity that may indicate fraud. Banks and payment networks use these systems to monitor transactions in real time, helping detect suspicious behavior before losses occur.

Medical research is another area experiencing transformation. Artificial intelligence systems can analyze biological datasets containing millions of data points, accelerating the search for new treatments and drug candidates.

These examples illustrate a broader trend.

AI is not simply replacing human workers. Instead, it is reshaping how work is performed. Many professionals are beginning to use AI as a productivity multiplier, allowing them to accomplish far more than would have been possible in earlier technological eras.

Jobs Most Likely to Change First

Not all jobs will be affected equally by artificial intelligence.

Certain professions involve highly structured tasks that follow predictable patterns. These types of tasks are particularly well suited for automation.

Examples include:

Data analysis

Document processing

Customer service interactions

Basic programming tasks

Administrative support work

In many cases, AI systems are already being integrated into these roles as productivity tools.

For example, legal professionals often spend large amounts of time reviewing documents during litigation or regulatory investigations.

Machine learning systems can assist by scanning thousands of documents and identifying relevant passages far more quickly than human reviewers.

Similarly, customer service departments are increasingly using AI chat systems capable of answering common questions automatically.

These systems allow companies to handle large volumes of inquiries without requiring equally large support teams.

However, the impact of AI extends far beyond these early use cases.

The Creation of New Professions

Although technological revolutions often eliminate certain jobs, they also tend to create entirely new professions.

During the early days of the internet, many people feared that automation would reduce employment opportunities.

Instead, the internet gave rise to new industries such as:

Search engine optimization

Digital marketing

Social media management

Mobile app development

Cybersecurity

AI is likely to produce similar effects.

New job categories are already beginning to emerge, including roles such as:

AI trainers who help refine machine learning systems

Prompt engineers who specialize in interacting effectively with AI models

AI safety researchers who study system reliability and alignment

Human-AI interaction designers who create interfaces for intelligent systems

As AI continues to evolve, additional professions will likely emerge that we cannot yet fully anticipate.

The Changing Nature of Expertise

One of the most interesting aspects of artificial intelligence is how it may change the concept of expertise.

In traditional professions, expertise often requires years of training and experience.

Doctors study medicine for more than a decade. Engineers spend years mastering mathematics and physics. Lawyers undergo extensive legal education and training.

AI systems are beginning to act as powerful support tools for these professions.

Medical AI systems can assist doctors by analyzing diagnostic images and identifying patterns that may indicate disease.

Engineering software can simulate complex designs and test structural integrity before physical prototypes are built.

Legal AI tools can analyze large collections of case law and identify relevant precedents.

These systems do not replace human expertise.

But they can augment it.

A skilled professional working with advanced AI tools may be able to accomplish far more than would have been possible in earlier eras.

Productivity and Economic Growth

The productivity gains associated with artificial intelligence could have profound economic consequences.

Productivity measures how efficiently an economy produces goods and services.

When productivity increases, societies often experience higher living standards, economic growth, and improved quality of life.

Throughout history, major technological innovations have driven productivity improvements.

Steam engines transformed manufacturing. Electricity revolutionized industry. Computers accelerated information processing.

Artificial intelligence may produce another major leap.

By automating routine tasks and accelerating complex analysis, AI could allow workers to focus on higher-level problem solving and creative work.

This shift may increase the overall productivity of the global economy.

Some economists believe that AI could contribute trillions of dollars in additional economic output over the coming decades.

However, these benefits may not be distributed evenly.

The Transition Challenge

While technological revolutions create long-term economic growth, the transition period can be disruptive.

When mechanized factories replaced manual textile production during the industrial revolution, many skilled workers lost their livelihoods.

Similarly, automation in manufacturing during the twentieth century displaced large numbers of industrial workers.

Artificial intelligence may produce similar disruptions in certain sectors.

Workers whose jobs involve highly repetitive cognitive tasks may face increased competition from automated systems.

This raises important questions for policymakers, educators, and business leaders.

How should societies prepare workers for an AI-driven economy?

What types of education and training will be most valuable in the future?

And how can the economic benefits of automation be shared broadly across society?

These questions will likely shape public policy debates in the coming decades.

Education in the Age of AI

As artificial intelligence becomes more integrated into the workplace, education systems may need to adapt.

Traditional education models often focus on memorization and routine problem-solving.

But these are precisely the types of tasks that AI systems perform extremely well.

The skills most valuable in an AI-driven world may instead include:

Critical thinking

Creativity

Complex problem solving

Interdisciplinary knowledge

Human communication and leadership

Education may increasingly emphasize these capabilities.

Students may also learn how to work effectively with AI tools, using them as collaborative partners in research, design, and innovation.

In this sense, AI may transform not only the workplace but also the process of learning itself.

Human Work in an AI World

Despite widespread concern about automation, it is unlikely that artificial intelligence will eliminate the need for human work entirely.

Human societies are constantly evolving.

New industries emerge as technologies create opportunities that did not previously exist.

In the long term, AI may shift human work toward areas where uniquely human capabilities remain essential.

These may include:

Creative expression

Strategic decision-making

Leadership and organization

Complex interpersonal communication

Ethical judgment

Artificial intelligence may handle many analytical and computational tasks, but humans will continue to play a critical role in shaping how these technologies are used.

The future of work may therefore involve collaboration rather than replacement.

Humans and intelligent machines working together to achieve goals that neither could accomplish alone.

The Beginning of a New Economic Era

Artificial intelligence represents more than a technological innovation.

It represents the beginning of a new economic era.

Just as previous industrial revolutions transformed agriculture, manufacturing, and transportation, AI may reshape knowledge work, research, and digital industries.

The transition will not happen overnight.

But the foundations of this transformation are already visible.

AI systems are becoming more capable.

Businesses are integrating machine learning into their operations.

Workers are learning how to collaborate with intelligent tools.

Over time, these changes may redefine the nature of work itself.

And the societies that adapt most effectively to this transformation may gain significant advantages in the emerging AI-driven economy.

Key Takeaways

• Artificial intelligence may automate certain cognitive tasks previously performed by humans.

• Many professions are already integrating AI tools to increase productivity.

• Some jobs will change significantly, while new professions will emerge.

• Education systems may need to adapt to prepare individuals for an AI-driven economy.

• Human-AI collaboration may become a defining feature of future workplaces.

Expert Insight

Technological revolutions rarely eliminate work entirely, but they often transform the types of work that societies perform. During the industrial revolution, machines replaced manual labor in many industries while creating new opportunities in manufacturing, engineering, and transportation.

Artificial intelligence may produce a similar transformation.

Rather than replacing human workers completely, AI systems may handle routine analytical tasks while humans focus on creativity, leadership, and strategic decision-making.

The most successful professionals in the future may be those who learn how to effectively collaborate with intelligent machines.

Strategic Question

If artificial intelligence can automate certain forms of cognitive work, what skills will become most valuable in the future workforce?

Will education systems adapt quickly enough to prepare individuals for a world in which humans collaborate with intelligent machines rather than compete against them?

Looking Ahead

With AI tools reshaping productivity, entrepreneurs are beginning to explore new business opportunities.

The next chapter examines the rapid growth of AI startups and the emergence of a new innovation ecosystem.

Chapter 6

The Artificial Intelligence Economic Explosion

When a technology multiplies human productivity, entire industries can appear almost overnight.

A New Engine of Global Growth

Every major technological revolution in history has reshaped the global economy.

The steam engine powered the industrial revolution and transformed manufacturing. Electricity enabled mass production and modern infrastructure. The internet created entirely new digital industries and connected billions of people across the globe.

Artificial intelligence may represent the next great economic engine.

But unlike earlier technologies that transformed specific industries, AI has the potential to influence nearly every sector of the global economy simultaneously.

Machine intelligence can analyze data, generate ideas, automate workflows, and assist with complex decision-making. These capabilities can increase productivity across industries ranging from healthcare and finance to manufacturing, logistics, and scientific research.

When productivity increases across many sectors at once, economic growth can accelerate dramatically.

For this reason, many economists believe that artificial intelligence may trigger one of the largest economic expansions in modern history.

From the Digital Economy to the Intelligence Economy

During the late twentieth century, the rise of the internet created what became known as the digital economy.

Businesses moved online. Information became widely accessible. Entire industries emerged around software, e-commerce, and digital communication.

Artificial intelligence may represent the next stage of this transformation.

Rather than simply digitizing information, AI introduces the ability to analyze and interpret that information at enormous scale.

In other words, the economy is moving from a system built on digital data to one built on digital intelligence.

Companies are beginning to integrate machine learning systems into nearly every aspect of their operations.

AI systems help analyze customer behavior, forecast demand, design products, optimize logistics networks, and detect financial risk.

As these capabilities spread, the structure of the global economy may begin shifting toward what some analysts describe as the intelligence economy.

In such a system, the most valuable companies may not simply control information.

They may control the systems capable of understanding and applying that information.

The Productivity Multiplier

One of the most powerful economic effects of artificial intelligence is its potential to act as a productivity multiplier.

Productivity measures how efficiently workers and organizations produce goods and services.

When productivity increases, economies can generate more value using the same amount of labor and resources.

Throughout history, technological breakthroughs have often produced major productivity gains.

Factories powered by steam engines allowed a small number of workers to produce goods at industrial scale. Electricity enabled mass production systems that dramatically increased manufacturing output. Computers accelerated information processing and administrative work.

Artificial intelligence may produce a similar leap.

AI systems can perform tasks that once required hours of human effort in a matter of seconds.

A marketing team can generate advertising concepts almost instantly using AI tools. Engineers can simulate complex designs before building physical prototypes. Researchers can analyze enormous scientific datasets in minutes rather than months.

When these productivity improvements occur across entire industries, the cumulative economic effect can be enormous.

The Trillion-Dollar Opportunity

Because artificial intelligence has applications across so many sectors, its economic impact could be vast.

Some economic forecasts estimate that AI could contribute tens of trillions of dollars to the global economy over the coming decades.

These gains may arise from several sources.

First, companies may reduce costs by automating routine processes.

Second, organizations may increase revenue by using AI to develop new products and services.

Third, entirely new industries may emerge around intelligent software systems.

The internet created companies that did not exist thirty years ago. Search engines, social media platforms, cloud computing providers, and online marketplaces became central components of the modern economy.

Artificial intelligence may produce a similar wave of new industries.

AI-driven robotics, autonomous logistics networks, intelligent medical research platforms, and advanced scientific discovery systems may all become major sectors of the future economy.

The companies that develop these technologies could grow into the economic giants of the twenty-first century.

The Companies Powering the AI Economy

The emerging AI economy is supported by several layers of technology companies.

At the foundation are semiconductor manufacturers and hardware developers producing the specialized processors required to train machine learning models.

Above this layer are cloud computing providers that operate large networks of data centers capable of running AI workloads at scale.

Another layer consists of companies developing foundational AI models, large systems trained on vast datasets that can perform a wide variety of tasks such as generating text, analyzing images, or writing software.

Finally, thousands of startups are building applications on top of this infrastructure, applying artificial intelligence to industries such as healthcare, finance, education, logistics, and entertainment.

Together, these layers form the economic ecosystem of the AI era.

Entrepreneurs in the Intelligence Economy

AI is not only reshaping existing corporations.

It is also creating new opportunities for entrepreneurs.

Historically, building large technology companies required significant resources. Entrepreneurs needed teams of engineers, expensive computing infrastructure, and years of development before launching a product.

Today, AI tools are reducing many of these barriers.

Small teams can use machine learning models to generate code, design software interfaces, analyze markets, and automate customer interactions.

As a result, individuals and small startups can now build products that once required large organizations.

This shift may lead to a surge of entrepreneurial experimentation.

Many of these experiments will fail, as is common during periods of rapid innovation. But some will evolve into major technology companies that shape the global economy.

Just as the early internet produced companies that now dominate digital commerce and communication, the AI era may produce a new generation of influential firms.

Economic Winners and Losers

Every technological revolution creates both opportunities and disruptions.

While artificial intelligence may increase productivity and create new industries, it may also transform existing job markets and business models.

Companies that successfully integrate AI into their operations may gain significant advantages over competitors that fail to adapt.

Similarly, workers who learn to collaborate effectively with AI tools may become more productive and valuable in the workforce.

At the same time, certain industries may face disruption as automation replaces routine tasks.

Navigating this transition will require thoughtful strategies from governments, businesses, and educational institutions.

The long-term economic benefits of AI may be enormous.

But the path toward that future may involve significant change.

The Early Stages of an Economic Transformation

Although artificial intelligence is advancing rapidly, the AI economy is still in its early stages.

Many of the most powerful applications of machine intelligence have not yet been fully developed.

New breakthroughs in robotics, biotechnology, scientific research, and energy systems may emerge as AI capabilities continue to expand.

The next decade may therefore represent only the beginning of a much larger economic transformation.

If artificial intelligence fulfills even a portion of its potential, it could become one of the most powerful engines of innovation and economic growth in human history.

Key Takeaways

• Artificial intelligence has the potential to increase productivity across nearly every sector of the global economy.

• The transition from a digital economy to an intelligence economy may reshape how businesses operate.

• AI systems can act as productivity multipliers, allowing individuals and organizations to accomplish more with fewer resources.

• New industries and companies are emerging around machine intelligence.

• The long-term economic impact of AI could reach tens of trillions of dollars

Expert Insight

Technological revolutions often produce waves of economic expansion that unfold over decades. The industrial revolution transformed manufacturing and transportation. The digital revolution reshaped communication, commerce, and information exchange.

Artificial intelligence may represent the next stage of this historical pattern.

By automating routine tasks, accelerating scientific discovery, and enabling new forms of innovation, AI could dramatically increase global productivity. The countries and companies that learn to harness these capabilities effectively may gain significant economic advantages in the emerging intelligence economy.

Strategic Question

If artificial intelligence dramatically increases productivity across the global economy, how will societies distribute the economic gains?

Will the benefits of the intelligence economy be broadly shared, or will they concentrate among a small number of technology companies and highly skilled workers?

Looking Ahead

While the economic potential of artificial intelligence is immense, its impact will extend far beyond markets and productivity.

The next chapter explores how intelligent machines may influence one of the most important technological frontiers of the coming decades: the development of advanced robotics and physical AI systems.

PART III

THE INTELLIGENCE ECONOMY

The Big Idea

Throughout history, economic systems have evolved alongside technological revolutions.

Agricultural societies were built around land and food production. The industrial revolution shifted economic power toward factories, machinery, and mass manufacturing. The digital revolution created an economy centered on information, software, and global networks.

Artificial intelligence may signal the emergence of the next stage in this progression.

In the coming decades, the most valuable economic resource may no longer be physical materials or raw information. Instead, it may be the ability to generate insight, analyze complex systems, and automate decision-making.

In other words, intelligence itself may become the most valuable form of capital.

Artificial intelligence expands access to this resource.

Machine learning systems can analyze enormous datasets, identify patterns within complex environments, and assist with solving problems that once required extensive human expertise. As these systems become more capable, they may dramatically increase the productive potential of individuals, organizations, and entire economies.

This shift suggests that the world may be entering a new economic phase , what can be described as the Intelligence Economy.

In this emerging system, companies capable of developing powerful AI systems may gain extraordinary influence. Entrepreneurs who learn to build products around machine intelligence may create entirely new industries. Nations that invest in AI infrastructure, research ecosystems, and talent development may shape the future balance of global economic power.

The implications extend far beyond the technology sector.

Artificial intelligence may transform how businesses are created, how wealth is generated, and how innovation spreads across society.

The chapters in this section explore how the intelligence economy may develop.

They examine the surge of AI startups, the frameworks that may define this new economic system, the areas where wealth creation may concentrate, and the possibility that artificial intelligence could trigger one of the largest economic expansions in modern history.

Understanding the intelligence economy is essential for understanding the future.

Because the companies, entrepreneurs, and nations that master this new form of capital may define the next era of global prosperity.

Why This Matters Now

AI is often discussed as a technological breakthrough.

But its most significant impact may be economic.

Throughout history, transformative technologies have not only changed how societies function; they have reshaped the structure of global economies. The industrial revolution created manufacturing empires. The digital revolution produced technology platforms that now dominate global markets.

Artificial intelligence may trigger the next economic transformation.

Machine intelligence is beginning to influence productivity, innovation, and the creation of entirely new industries. Businesses are integrating AI systems into research, product development, logistics, and financial decision-making. Entrepreneurs are launching companies built around intelligent software tools. Governments are investing heavily in the infrastructure required to support large-scale AI development.

These changes suggest that the world may be entering a new economic phase.

In this emerging system, what we might call the intelligence economy, the ability to generate insight, analyze data, and automate complex tasks becomes one of the most valuable forms of capital.

Understanding this transformation is essential.

The chapters in this section explore how artificial intelligence may reshape economic structures, where new wealth may be created, and why the global race for AI leadership may determine which nations and companies define the next era of technological progress.

"AI will reshape every industry and create enormous economic opportunity."

—Andrew Ng

Chapter 7

The Artificial Intelligence Startup Explosion

Throughout history, the greatest economic opportunities have appeared during technological revolutions. Artificial intelligence may create the largest wave of startups the world has ever seen.

A New Wave of Innovation

Major technological revolutions rarely emerge quietly.

They tend to trigger waves of innovation that reshape entire industries. Entrepreneurs begin experimenting with new ideas, investors search for opportunities, and startups emerge to challenge established companies.

AI is now entering this phase.

The rapid progress of machine learning technologies has sparked what many analysts describe as an AI startup explosion. Across the world, entrepreneurs are building companies focused on applying artificial intelligence to nearly every sector of the economy.

Healthcare startups are using AI to accelerate drug discovery. Financial technology companies are using machine learning to detect fraud and analyze investment opportunities. Manufacturing firms are integrating intelligent automation into production systems.

In short, artificial intelligence is becoming a foundational technology for a new generation of businesses.

And like previous technological revolutions, the companies that emerge during this period may define the global economy for decades to come.

Lessons from Previous Technology Waves

To understand the significance of the AI startup boom, it is helpful to look at earlier technological revolutions.

The rise of the internet during the 1990s created entirely new industries. At the time, many people viewed the internet primarily as a communication tool. Few could have predicted the scale of the companies that would emerge from it.

Yet over time, internet-based startups grew into some of the most powerful corporations in the world.

Online marketplaces transformed retail. Search engines reshaped how people access information. Social networks changed how individuals communicate and share content.

These companies did not simply adapt existing industries.

They created entirely new economic ecosystems.

Artificial intelligence may follow a similar trajectory.

The most successful AI startups may not simply improve existing software tools. They may invent new categories of technology that fundamentally change how work is performed.

The Lowering of Barriers to Entry

One of the reasons the AI startup ecosystem has grown so rapidly is the decreasing cost of building software products.

In earlier decades, launching a technology company required significant resources. Developers needed access to expensive computing hardware, large software teams, and specialized infrastructure.

Today, many of these barriers have been reduced.

Cloud computing platforms provide access to powerful computing resources without requiring companies to build their own data centers. Open-source machine learning frameworks allow developers to build sophisticated AI systems without starting from scratch.

Large language models and other AI tools can assist with tasks such as writing code, generating documentation, and designing user interfaces.

As a result, small teams of engineers can now build products that previously required large organizations.

This shift has opened the door for a new generation of entrepreneurs.

AI as a Platform Technology

AI is increasingly being viewed as a platform technology.

A platform technology is one that supports many different applications across multiple industries.

Electricity, for example, became a platform for countless innovations ranging from household appliances to industrial machinery.

The internet became a platform for communication, commerce, and digital media.

Artificial intelligence may become a platform for intelligent software systems capable of performing complex tasks across many sectors.

This platform effect is one of the reasons investors have become increasingly interested in AI startups.

Rather than focusing on a single application, AI technologies can often be adapted to many different uses.

A machine learning model designed for analyzing customer behavior might also be applied to financial risk assessment, healthcare diagnostics, or supply chain optimization.

This flexibility makes AI technologies extremely valuable as building blocks for new businesses.

Venture Capital and the AI Investment Boom

As artificial intelligence technologies have matured, venture capital firms have dramatically increased their investments in AI startups.

Investors view AI as one of the most transformative technologies of the coming decades. They are therefore eager to support companies that may become leaders in the emerging AI economy.

Funding for AI startups has grown rapidly in recent years.

Entrepreneurs developing new machine learning tools, AI infrastructure platforms, robotics systems, and data analytics software have raised billions of dollars in investment capital.

Some of these companies are focused on developing foundational technologies such as AI chips, data infrastructure platforms, or model training systems.

Others are building specialized applications designed for particular industries.

The result is a rapidly expanding ecosystem of startups exploring how artificial intelligence can be applied to real-world problems.

Industries Being Reinvented

Artificial intelligence startups are emerging across many industries.

In healthcare, AI companies are developing tools that analyze medical images, assist with diagnosis, and accelerate drug discovery. Machine learning systems can examine vast biological datasets and identify patterns that might take human researchers years to uncover.

In finance, AI startups are building systems capable of analyzing financial markets, detecting fraud, and automating risk assessment.

In manufacturing, intelligent robotics systems are improving production efficiency and reducing operational costs.

Transportation companies are experimenting with autonomous driving technologies, logistics optimization systems, and AI-powered traffic management.

Even creative industries such as music, film, and design are being influenced by AI-driven tools that assist with content generation.

The range of applications continues to grow.

The Rise of AI Infrastructure Startups

While many AI startups focus on applications, another category is emerging that focuses on infrastructure.

These companies are developing the tools and platforms required to build, train, and deploy machine learning models.

Examples include:

Data management platforms for training datasets

Model deployment systems for running AI applications in production

AI chip manufacturers designing specialized hardware for machine learning

Software frameworks for managing large-scale training environments

These infrastructure companies play a critical role in the AI ecosystem.

Just as cloud computing platforms helped enable the growth of internet startups, AI infrastructure companies are helping accelerate the development of machine learning applications.

Competition with Technology Giants

The rise of AI startups has created both opportunities and challenges.

Large technology companies already possess many advantages in artificial intelligence development.

They control vast computing infrastructure, large datasets, and experienced engineering teams.

However, startups often possess advantages of their own.

Smaller companies can move quickly, experiment with new ideas, and focus on niche applications that may not initially attract the attention of larger corporations.

Many successful technology companies began as small startups competing against larger rivals.

The AI industry may follow a similar pattern.

Some of today's small AI startups could grow into the technology giants of tomorrow.

The Entrepreneurial Opportunity

For entrepreneurs, the rise of artificial intelligence represents an extraordinary opportunity.

The technology is still evolving rapidly, and many potential applications have yet to be explored.

Individuals with technical expertise, industry knowledge, or creative ideas may find opportunities to build new products and services powered by AI.

The key challenge is identifying problems that artificial intelligence can solve more effectively than existing tools.

Some of the most successful AI companies may emerge from industries that have not traditionally been associated with advanced technology.

Agriculture, logistics, energy management, and environmental monitoring are just a few areas where AI-driven innovation may produce new business opportunities.

As the technology continues to mature, the scope of these opportunities will likely expand.

The Early Days of an AI Economy

The rapid growth of AI startups suggests that the world may be entering the early stages of a new technological economy.

Just as the internet created an ecosystem of companies focused on digital communication, e-commerce, and online media, artificial intelligence may give rise to a new generation of intelligent software platforms.

Some of these companies will fail, as is common during periods of rapid innovation.

But others may grow into influential organizations that shape the future of technology.

The AI startup explosion is therefore not merely a trend.

It is a sign that artificial intelligence is transitioning from research laboratories into the broader economy.

Entrepreneurs are beginning to experiment with its possibilities.

Investors are searching for the next generation of technology leaders.

And the companies being built today may help define the structure of the global economy in the decades ahead.

Key Takeaways

• AI is fueling a surge of startup activity across many industries.

• Entrepreneurs are applying AI technologies to healthcare, finance, logistics, and manufacturing.

• Cloud computing and open-source tools have lowered the barriers to launching technology companies.

• Venture capital investment in AI startups has grown rapidly.

• The AI startup ecosystem may produce the next generation of global technology leaders.

Expert Insight

Periods of technological transformation often produce waves of entrepreneurial experimentation. During the early days of the internet, thousands of startups explored new ways to use digital communication and online commerce.

Many of these companies failed, but others evolved into the major technology platforms that shape today's digital economy.

Artificial intelligence may produce a similar cycle of experimentation and innovation. As machine learning tools become more accessible, entrepreneurs around the world are developing new products and services powered by intelligent software.

The companies that succeed in this environment may help define the structure of the AI economy.

Strategic Question

If artificial intelligence lowers the barriers to launching new companies, will the next generation of global technology leaders emerge from small startup teams rather than established corporations?

What new industries might appear when entrepreneurs gain access to powerful AI tools?

Looking Ahead

While startups are exploring new applications of artificial intelligence, entire industries are already being reshaped.

The next chapter examines how AI is transforming healthcare and medical research.

Chapter 8

The Intelligence Economy Framework

What happens when intelligence itself becomes a scalable resource? The answer may define the next economic era.

The Intelligence Economy Model

The transformation described in this book can be understood through a simple structural model.

The intelligence economy is composed of four interconnected layers:

Compute Infrastructure

The hardware foundation powering artificial intelligence, including semiconductors, GPUs, and global data centers.

Foundation Intelligence

Large-scale AI models capable of performing many tasks, including language generation, data analysis, and reasoning.

Industry Applications

Companies applying AI technologies to sectors such as healthcare, finance, manufacturing, logistics, and energy.

Human AI Collaboration

The interaction between human expertise and intelligent systems, creating new forms of productivity and innovation.

These layers form the architecture of the intelligence economy.

Just as the industrial economy relied on factories, transportation networks, and energy infrastructure, the intelligence economy depends on computing power, machine learning systems, and the collaboration between humans and intelligent machines.

Understanding this structure helps reveal where the greatest technological and economic transformations may occur.

A New Economic Model Emerges

Throughout history, economic systems have evolved alongside the technologies that power them.

Agricultural societies were built around land, labor, and food production. Industrial economies were driven by factories, machinery, and mass manufacturing. The digital era introduced software, data networks, and global information exchange.

AI is now giving rise to something different.

A new economic structure is beginning to emerge, one built around intelligence itself.

In this system, the ability to generate insight, analyze complex information, and make effective decisions becomes one of the most valuable economic resources in the world.

This transformation can be understood through what we may call the Intelligence Economy Framework.

Rather than focusing only on technology, this framework explains how artificial intelligence reshapes the structure of economic value.

From Information to Intelligence

During the early decades of the internet, the most valuable companies were those that controlled information.

Search engines organized the world's knowledge. Social media platforms aggregated communication between billions of people. E-commerce platforms managed global marketplaces of digital and physical goods.

Information became the currency of the digital economy.

Artificial intelligence introduces a new layer above that.

Instead of merely storing or distributing information, AI systems can interpret it.

They can identify patterns, generate insights, and propose decisions. In other words, they transform raw data into usable intelligence.

The companies capable of performing this transformation at scale may become the most powerful organizations of the next economic era.

Just as previous generations built companies around oil, electricity, or computing infrastructure, the next generation of companies may build their influence around machine intelligence.

The Four Layers of the Intelligence Economy

To understand how the AI economy operates, it is useful to examine the ecosystem through four interconnected layers.

Each layer contributes to the creation and application of artificial intelligence.

Together, they form the structural foundation of the intelligence economy.

Layer One: Compute Infrastructure

At the base of the intelligence economy lies computational power.

Training modern AI systems requires enormous amounts of processing capability. Specialized processors, high-speed networking hardware, and vast data centers perform trillions of calculations required to train machine learning models.

Semiconductor companies, cloud computing providers, and data center operators form the infrastructure layer of artificial intelligence.

Without this hardware foundation, advanced AI systems could not exist.

Control over computational infrastructure therefore represents one of the most strategic resources in the intelligence economy.

Layer Two: Foundation Models

Above the infrastructure layer are the organizations building large-scale AI models.

These systems are trained on massive datasets and designed to perform a wide range of tasks, including language generation, image recognition, and complex reasoning.

Because these models can be adapted to many different applications, they function as foundational technologies for the broader AI ecosystem.

In many ways, foundation models operate like operating systems for intelligence.

Developers build applications on top of them, businesses integrate them into workflows, and researchers adapt them to specialized tasks.

Layer Three: AI Applications

The third layer consists of companies applying artificial intelligence to specific industries.

These organizations use machine learning to build products that solve real-world problems.

Examples include AI systems that assist with medical diagnosis, automate financial analysis, optimize supply chains, or generate software code.

This layer represents the interface between artificial intelligence and everyday economic activity.

While infrastructure and foundation models enable the technology, applications determine how it affects industries, businesses, and consumers.

Layer Four: Human-AI Collaboration

The final layer of the intelligence economy involves the interaction between humans and intelligent systems.

Artificial intelligence rarely operates in isolation. Instead, it often functions as a tool that amplifies human capabilities.

Doctors may use AI systems to analyze medical images. Engineers may rely on machine learning simulations to design new materials. Entrepreneurs may use AI tools to develop products, analyze markets, and manage business operations.

In this layer, economic value emerges from collaboration between human expertise and machine intelligence.

Individuals capable of effectively integrating AI into their work may gain significant productivity advantages.

The Intelligence Multiplier

One of the most important concepts within the intelligence economy is what we might call the intelligence multiplier.

Artificial intelligence does not simply automate tasks.

It increases the effective cognitive capacity of individuals and organizations.

A scientist working with AI tools can analyze far more data than would be possible manually. A software developer assisted by AI coding systems can write and test programs more quickly. A startup team using AI tools can perform work that once required entire departments.

In effect, AI multiplies human intellectual output.

When this multiplier effect spreads across entire industries, the overall productive capacity of the economy increases dramatically.

A New Competitive Landscape

The emergence of the intelligence economy may reshape global competition.

During the industrial era, economic power was often determined by manufacturing capability and access to natural resources.

In the digital era, technological leadership and information networks became critical.

In the intelligence economy, the most competitive nations and organizations may be those that combine three capabilities:

Advanced computing infrastructure

Large datasets

Highly skilled technical talent

Countries and companies that successfully integrate these elements may lead the development of next-generation AI systems.

Those that fail to do so may struggle to remain competitive.

The Early Architecture of the Intelligence Economy

Although artificial intelligence is advancing rapidly, the intelligence economy is still in its early stages.

Many of the technologies that will define this system, autonomous robotics, advanced reasoning models, AI-assisted scientific discovery, are still developing.

But the underlying structure is beginning to take shape.

Infrastructure providers are expanding global computing networks. Researchers are building increasingly capable AI models. Entrepreneurs are launching startups that apply machine intelligence to new industries.

Together, these developments represent the first building blocks of a new economic architecture.

Just as the industrial revolution required factories and railroads, the intelligence economy requires algorithms, computing infrastructure, and human creativity working together.

The societies that learn to harness these elements most effectively may shape the future of the global economy.

THE INTELLIGENCE ECONOMY STACK (FRAMEWORK DIAGRAM)

Layer 5 — AI-Enabled Industries
Healthcare, finance, robotics, manufacturing, logistics, education, scientific research, media, and entertainment.

Layer 4 — AI Applications
AI copilots, enterprise AI tools, automation systems, AI research assistants, and intelligent business software.

Layer 3 — AI Platforms
Large language models, multimodal models, AI APIs, model training systems, and AI development platforms.

Layer 2 — AI Infrastructure
Cloud computing, data centers, AI supercomputers, high-performance networking systems.

Layer 1 — Semiconductor Foundations
AI chips, GPUs, specialized processors, advanced semiconductor manufacturing.

Understanding the Intelligence Economy Stack

While the intelligence economy can be described through four conceptual layers, it can also be visualized as a broader technological stack. Each layer builds upon the capabilities of the one beneath it.

At the foundation are semiconductor technologies that power modern computing systems. These chips enable the enormous computational capacity required to train machine learning models.

Above this layer lies the infrastructure that operates these systems, including global data centers, cloud computing platforms, and AI supercomputing clusters capable of processing massive datasets.

The next layer contains AI platforms and foundation models. These systems function as the core engines of machine intelligence, allowing developers and organizations to build intelligent software applications.

Above the platform layer sit the applications that bring artificial intelligence into everyday use. These include AI copilots, automation systems, research assistants, and business analytics tools.

At the top of the stack are the industries transformed by these technologies. Healthcare, finance, robotics, logistics, and scientific research are all beginning to integrate artificial intelligence into their operations.

Together, these layers form the technological and economic architecture of the intelligence economy.

Key Takeaways

• Artificial intelligence may transform the global economy by shifting value from information to intelligence.

• The intelligence economy consists of four layers: compute infrastructure, foundation models, AI applications, and human-AI collaboration.

• AI acts as an intelligence multiplier, increasing the productive capacity of individuals and organizations.

• Control over computing infrastructure, data, and technical talent may become critical sources of economic power.

• The intelligence economy is still in its early stages but may define the next era of global innovation.

Expert Insight

Economic history often unfolds through technological platforms that enable new waves of productivity. Electricity powered industrial manufacturing. The internet enabled global communication and digital commerce.

Artificial intelligence may become the platform technology of the intelligence economy.

As machine learning systems become more capable, they may transform not only how businesses operate but also how knowledge itself is created and applied.

Strategic Question

If intelligence becomes the most valuable economic resource of the twenty-first century, who will control it?

Will machine intelligence remain concentrated among a small number of corporations and governments, or will it become a widely accessible tool that empowers individuals and entrepreneurs around the world?

Looking Ahead

While artificial intelligence is reshaping the structure of the global economy, its most visible impact may occur in the physical world.

The next chapter explores how robotics and intelligent machines are beginning to transform industries that depend on physical labor and infrastructure.

Chapter 9

The Artificial Intelligence Wealth Map

In every technological revolution, wealth flows toward those who understand the shift first.

Where Value Is Created in the Age of Artificial Intelligence

Technological revolutions rarely distribute economic value evenly.

During the industrial revolution, the greatest wealth was created not only by factories themselves but by the industries that enabled and expanded industrial production: railroads, steel manufacturing, and energy infrastructure.

The digital revolution followed a similar pattern.

Early internet companies created new ways to communicate and exchange information, but the greatest economic value ultimately emerged in several key sectors: search engines, cloud computing, e-commerce platforms, and digital advertising networks.

Artificial intelligence may follow a comparable trajectory.

While the technology itself is remarkable, the largest economic gains may emerge in specific industries that successfully build, deploy, and scale intelligent systems.

Understanding where these opportunities exist is essential.

The AI Wealth Map provides a way to visualize where economic value may accumulate during the rise of machine intelligence.

The First Layer: AI Infrastructure

At the base of the AI wealth map lies infrastructure.

Artificial intelligence requires enormous computational power, advanced semiconductor technology, and vast data center networks capable of running large-scale machine learning systems.

These systems perform trillions of mathematical calculations every second.

Companies that design specialized processors, build cloud computing platforms, and operate large data centers form the foundation of the AI economy.

Historically, infrastructure companies often become some of the most valuable organizations during technological revolutions.

Railroad companies dominated transportation during the industrial era. Telecommunications networks expanded alongside the growth of the internet.

In the AI era, computing infrastructure may play a similar role.

Organizations capable of building the physical engines of machine intelligence may control one of the most critical resources of the twenty-first century: computational power.

The Second Layer: Foundation Intelligence

Above the infrastructure layer lies another important category of value creation.

Foundation intelligence.

These are the organizations developing large-scale artificial intelligence models capable of performing a wide variety of tasks.

Such systems can generate language, analyze images, assist with software development, and process enormous datasets.

Because these models can serve many industries simultaneously, they function as a technological platform upon which other companies can build products and services.

In economic terms, foundation models behave much like operating systems.

Just as mobile operating systems enabled entire ecosystems of smartphone applications, foundational AI systems may support a vast ecosystem of intelligent software products.

Companies capable of developing and maintaining these systems may therefore occupy an influential position within the AI economy.

The Third Layer: Industry Applications

While infrastructure and foundational models enable artificial intelligence, the largest economic transformation may occur in the industries that apply AI to real-world problems.

Across the global economy, organizations are beginning to integrate machine learning into their operations.

Healthcare companies are using AI to accelerate drug discovery and assist with medical diagnosis.

Financial institutions are deploying machine learning systems to detect fraud and analyze complex market patterns.

Manufacturing firms are integrating intelligent robotics and predictive maintenance systems into production lines.

Logistics companies are optimizing supply chains and transportation networks using AI-driven forecasting models.

These applications illustrate how machine intelligence can increase efficiency and unlock new capabilities across many sectors.

As adoption spreads, entire industries may be reshaped.

The Fourth Layer: Autonomous Systems

AI is not limited to digital software systems.

Increasingly, machine intelligence is being integrated into physical machines capable of interacting with the real-world.

Autonomous vehicles, intelligent manufacturing robots, warehouse automation systems, and AI-powered drones represent early examples of this trend.

These systems combine software intelligence with physical engineering.

The result is a new generation of machines capable of performing tasks that once required human labor.

As robotics technology advances, autonomous systems may become a major economic sector.

Entire industries, from logistics and agriculture to construction and transportation, may adopt intelligent machines that operate with increasing independence.

This shift could significantly expand the economic reach of artificial intelligence beyond the digital realm.

The Fifth Layer: Human Intelligence Amplification

The final layer of the AI wealth map involves a less visible but equally powerful transformation.

Artificial intelligence may dramatically amplify human productivity.

Instead of replacing human expertise entirely, many AI systems function as tools that enhance the capabilities of individuals.

A scientist working with machine learning tools can analyze far larger datasets than would be possible manually.

A software engineer assisted by AI coding systems can develop applications more quickly.

An entrepreneur using AI-driven analytics tools can evaluate markets and design products with unprecedented speed.

In each case, the individual remains central to the process.

But their capabilities are expanded through collaboration with intelligent machines.

This amplification of human productivity may become one of the most important drivers of economic growth in the intelligence economy.

Where the Largest Opportunities May Appear

Although artificial intelligence will influence many industries, certain sectors may experience particularly dramatic growth.

These sectors often share three characteristics.

They involve complex data analysis.

They benefit from automation or optimization.

And they operate at large economic scale.

Examples include healthcare, biotechnology, financial services, energy systems, logistics networks, and advanced manufacturing.

Entrepreneurs and investors are therefore closely watching these industries for opportunities to apply machine intelligence.

Some of the most influential companies of the next several decades may emerge from these areas.

The Uneven Distribution of AI Wealth

While artificial intelligence may increase global economic productivity, the distribution of wealth generated by this technology may not be uniform.

Certain companies may capture enormous value by controlling key infrastructure, foundational models, or widely adopted platforms.

Meanwhile, other organizations may adopt AI primarily as a productivity tool rather than as a source of large-scale revenue growth.

This uneven distribution has historical precedent.

During the digital revolution, a relatively small number of technology companies became extraordinarily influential while many other firms simply incorporated digital tools into their existing operations.

Artificial intelligence may produce a similar pattern.

Understanding where value accumulates within the AI wealth map may therefore become an important strategic question for entrepreneurs, investors, and policymakers.

The Map Is Still Being Drawn

Despite the rapid progress of artificial intelligence, the AI wealth map remains incomplete.

Many of the most powerful applications of machine intelligence have not yet been fully developed.

New breakthroughs in robotics, biotechnology, materials science, and energy systems may open entirely new sectors of economic opportunity.

In this sense, the intelligence economy is still being built.

The companies, institutions, and individuals shaping the development of artificial intelligence today may influence the economic landscape of the twenty-first century.

And the map of where value will ultimately accumulate is still being drawn.

AI WEALTH MAP ECONOMIC OPPORTUNITY IN THE INTELLIGENCE ECONOMY

Layer 5 — Human Intelligence Amplification
AI copilots, research assistants, creative tools, decision-support systems

Layer 4 — Autonomous Systems
Robotics, autonomous vehicles, drones, intelligent manufacturing systems

Layer 3 — Industry Applications
Healthcare AI, fintech AI, logistics optimization, energy systems, smart infrastructure

Layer 2 — Foundation Intelligence
Large language models, multimodal AI systems, reasoning models, AI platforms

Layer 1 — AI Infrastructure
Semiconductors, GPUs, AI chips, cloud computing, global data centers

Visualizing the AI Wealth Map

The AI wealth map illustrates how economic value is created across multiple layers of the intelligence economy. Each layer represents a different category of companies, technologies, and opportunities.

At the base lies infrastructure: semiconductor manufacturers, chip designers, cloud providers, and data center operators. These organizations build the physical computing systems that power machine intelligence.

Above infrastructure are foundation intelligence systems, large AI models that function as general-purpose engines for reasoning, language generation, and data analysis. These models allow developers to build thousands of applications.

The next layer includes industry applications. Companies across healthcare, finance, manufacturing, logistics, and energy are integrating AI systems into their operations, unlocking new efficiencies and capabilities.

Above these applications sit autonomous systems, including robots, drones, and intelligent machines that combine AI software with physical engineering to perform real-world tasks.

At the top of the map is human intelligence amplification. AI tools increasingly act as collaborators that enhance human productivity. Scientists, engineers, entrepreneurs, and professionals use intelligent systems to analyze data, design solutions, and generate ideas faster than ever before.

Together, these layers form a map of where economic value may accumulate during the rise of artificial intelligence.

Key Takeaways

• Artificial intelligence may generate economic value across several interconnected layers of technology and industry.

• The AI wealth map includes infrastructure, foundation intelligence, industry applications, autonomous systems, and human-AI collaboration.

• Companies controlling critical infrastructure or widely used AI platforms may capture significant economic value.

• Many industries may experience productivity gains and innovation driven by machine intelligence.

• The distribution of AI-generated wealth may concentrate among organizations capable of scaling intelligent technologies.

Expert Insight

Technological revolutions often create new economic hierarchies.

The companies that build foundational infrastructure and widely adopted platforms frequently capture a disproportionate share of value. Railroads and energy companies dominated the industrial era, while digital platforms and cloud providers became central to the internet economy.

Artificial intelligence may produce a similar pattern.

Organizations that control key layers of the AI ecosystem, including computing infrastructure, foundational models, and scalable applications, may emerge as some of the most influential economic actors of the coming decades.

Strategic Question

If artificial intelligence reshapes the global economy, which organizations will control the most valuable layers of the AI wealth map?

Will economic power concentrate among a small number of technology platforms, or will new entrepreneurs emerge to challenge established leaders?

Looking Ahead

While artificial intelligence is already transforming industries and economic structures, its long-term implications extend even further.

The next chapter explores how the development of advanced artificial intelligence may eventually lead toward systems capable of reasoning and learning across many domains, what researchers describe as artificial general intelligence.

Chapter 10

The $100 Trillion AI Race

The next economic race may not be for territory or resources, but for intelligence itself.

The Largest Economic Expansion in History

Throughout human history, technological revolutions have repeatedly reshaped the scale of the global economy.

The agricultural revolution allowed civilizations to support large populations. The industrial revolution dramatically expanded manufacturing output and transportation networks. The digital revolution connected billions of people through information systems and created entirely new industries.

Artificial intelligence may represent the next stage of this economic transformation.

But the scale of its potential impact may exceed that of many previous technological shifts.

Some analysts now estimate that artificial intelligence could contribute tens of trillions of dollars to global economic activity during the coming decades. When combined with the secondary effects of productivity growth, new industries, and scientific discovery, the total economic expansion associated with machine intelligence could reach levels rarely seen in history.

In other words, humanity may be entering what could become one of the largest economic expansions ever recorded.

Why Artificial Intelligence Scales So Quickly

One of the defining characteristics of artificial intelligence is its ability to scale rapidly.

Traditional industries often expand slowly because they require physical infrastructure, large workforces, and extensive supply chains. Building factories, constructing transportation systems, and manufacturing physical goods all require time and capital.

Artificial intelligence operates differently.

Once an AI system has been developed and trained, it can often be deployed across millions of users almost instantly through digital networks.

A software system capable of analyzing data, generating text, or assisting with decision-making can be accessed simultaneously by organizations around the world.

This scalability allows AI technologies to spread far more quickly than earlier industrial technologies.

As a result, the economic effects of artificial intelligence may unfold at unprecedented speed.

The Productivity Shock

One of the primary drivers of the AI economic expansion is productivity.

Productivity measures how efficiently an economy produces goods and services. When productivity increases, societies can generate greater economic output using the same amount of labor and resources.

Artificial intelligence has the potential to dramatically increase productivity across many sectors.

Researchers can analyze scientific data more quickly. Engineers can design and simulate complex systems with the assistance of machine learning tools. Businesses can automate large portions of administrative work.

In each of these cases, individuals become capable of accomplishing far more work than before.

If this productivity multiplier spreads across the global economy, the cumulative effect could be enormous.

Entire industries may become more efficient, more innovative, and more capable of solving complex challenges.

The Industries Driving the AI Expansion

Although artificial intelligence may influence nearly every sector of the economy, several industries may play particularly significant roles in driving the AI expansion.

Healthcare and biotechnology represent one of the most promising areas.

Machine learning systems are already assisting researchers in analyzing biological data, predicting protein structures, and identifying potential drug candidates. These capabilities may accelerate the development of new medical treatments and therapies.

Energy systems represent another major opportunity.

Artificial intelligence can help optimize power grids, improve renewable energy forecasting, and design more efficient energy storage technologies.

Transportation and logistics are also undergoing transformation.

AI-driven systems can optimize shipping routes, coordinate supply chains, and support the development of autonomous vehicles capable of operating with minimal human intervention.

Manufacturing may experience a new wave of automation through intelligent robotics and predictive maintenance systems.

Even scientific research itself may accelerate as AI systems assist researchers in analyzing complex datasets and generating new hypotheses.

Together, these industries could form the backbone of the emerging AI economy.

The Rise of Trillion-Dollar Industries

Technological revolutions often create entirely new categories of economic activity.

The internet produced global industries such as online commerce, digital advertising, cloud computing, and social media platforms.

Artificial intelligence may generate its own set of transformative industries.

AI-driven robotics could reshape manufacturing and logistics. Intelligent healthcare platforms may accelerate drug discovery and personalized medicine. Autonomous transportation networks could transform the movement of goods and people.

Advanced materials discovery powered by machine learning could enable new technologies in fields ranging from aerospace engineering to renewable energy.

These emerging sectors may give rise to companies of unprecedented scale.

Some analysts believe that several industries influenced by artificial intelligence could eventually reach multi-trillion-dollar valuations.

While predicting specific outcomes is difficult, the magnitude of the opportunity is clear.

The Global Competition for the AI Future

Because artificial intelligence may drive enormous economic value, governments and corporations around the world are racing to lead its development.

Countries that successfully build strong AI ecosystems may gain significant economic advantages.

These ecosystems often include research universities, advanced semiconductor manufacturing, large computing infrastructure, and vibrant startup communities.

Nations capable of combining these elements may become leaders in the intelligence economy.

Meanwhile, technology companies are investing billions of dollars in AI research, infrastructure, and talent acquisition.

This competition is not limited to economic growth alone.

Artificial intelligence may influence national security, scientific leadership, and global technological influence.

For these reasons, the AI race has become one of the defining strategic competitions of the twenty-first century.

Opportunities for Entrepreneurs

Although large technology companies currently dominate many aspects of artificial intelligence development, the AI revolution is still in its early stages.

Entrepreneurs continue to explore new ways of applying machine intelligence to real-world challenges.

Startups are developing AI-powered tools for healthcare diagnostics, financial analytics, software development, education, and environmental monitoring.

Cloud computing platforms and open-source machine learning frameworks have lowered the barriers to launching new technology companies.

As a result, small teams of engineers and researchers are now capable of building products that once required large organizations.

Many of the most influential companies of the next generation may emerge from this wave of entrepreneurial experimentation.

The Risks of Rapid Transformation

While the economic potential of artificial intelligence is enormous, rapid technological change also introduces uncertainty.

Industries may face disruption as automation reshapes traditional business models.

Workforces may need to adapt to new technologies and evolving skill requirements.

Governments may face difficult questions about regulation, privacy, and the responsible development of advanced AI systems.

Managing these challenges will require careful collaboration between policymakers, businesses, and researchers.

The long-term benefits of artificial intelligence may be profound.

But navigating the transition toward the intelligence economy will require thoughtful leadership.

Standing at the Beginning

Despite the rapid progress of artificial intelligence, humanity is still at the early stages of this transformation.

Many of the technologies that will define the AI era, including advanced robotics, scientific discovery systems, and autonomous infrastructure, are only beginning to emerge.

The coming decades may therefore represent a period of extraordinary innovation.

If artificial intelligence fulfills even a portion of its potential, the global economy may experience one of the most significant expansions in human history.

The economic transformation driven by AI is unlikely to be a single event or a single number.

It may be the beginning of a new technological era, one defined by intelligence itself.

THE $100 TRILLION AI ECONOMY MODEL

Scientific Discovery

AI-assisted research

Drug discovery

Advanced materials

Energy innovation Industry Transformation

Healthcare

Finance

Manufacturing

Transportation

Energy systems Enterprise Productivity

Automation of workflows

AI copilots

Decision-support systems

Business intelligence Digital Intelligence Platforms

Large language models

Multimodal AI systems

AI development platforms

Cloud AI services Infrastructure Expansion

Semiconductors

AI chips

Data centers

Global computing networks

Visualizing the AI Economic Expansion

The intelligence economy model illustrates how artificial intelligence may generate economic value across several interconnected layers with the scale of that value depending heavily on the pace of adoption, the depth of integration, and the policy environments that govern deployment.

At the base of the model lies infrastructure expansion. Semiconductor manufacturers, AI chip designers, cloud computing providers, and global data center networks supply the computational power required to train and operate machine learning systems.

Above infrastructure are digital intelligence platforms. Large language models and multimodal AI systems serve as general-purpose engines that enable developers and organizations to build intelligent applications.

The next layer represents enterprise productivity. AI copilots, automation systems, and data analytics tools allow organizations to operate more efficiently by accelerating research, development, and decision-making.

Above this layer sits industry transformation. AI is already reshaping sectors such as healthcare, finance, manufacturing, transportation, and energy systems.

At the top of the model is scientific discovery. Machine learning systems may accelerate breakthroughs in biotechnology, materials science, climate modeling, and energy innovation.

Together these layers illustrate why some economists believe artificial intelligence could generate one of the largest economic expansions in human history.

The Intelligence Economy Model

AI is not a single technology or industry. It is an ecosystem.

The rise of machine intelligence is creating a new economic structure built from several interconnected layers. These layers interact with one another to produce technological innovation, economic value, and new forms of productivity.

At the base of the system lies infrastructure. Above it are the models that generate machine intelligence. On top of those models sit applications and industries that apply AI to real-world challenges. Finally, at the highest level, humans collaborate with intelligent systems to amplify creativity, research, and decision-making.

Together, these layers form the architecture of the intelligence economy.

THE INTELLIGENCE ECONOMY

Human Intelligence Amplification
Scientists
Engineers
Entrepreneurs
Creators
Human–AI collaboration Autonomous & Intelligent Systems
Robotics
Autonomous vehicles
AI agents
Industrial automation Industry Transformation
Healthcare
Finance
Manufacturing
Energy
Logistics
Education AI Platforms & Foundation Models
Large language models
Multimodal AI systems
AI operating platforms
Machine learning ecosystems Compute Infrastructure
Semiconductors
AI chips
GPUs
Cloud computing
Data centers
Global computing networks

Explanation

How the Layers Interact Each layer of the intelligence economy supports the layers above it.

Compute infrastructure provides the computational power required to train and operate machine learning systems. Foundation models transform this computational capacity into usable intelligence capable of generating language, analyzing images, and reasoning about complex problems.

Developers and entrepreneurs build applications on top of these models, applying machine intelligence to industries such as healthcare, finance, logistics, and scientific research.

As these technologies mature, they enable autonomous systems capable of interacting with the physical world through robotics and automation.

At the highest level, humans collaborate with these intelligent tools. Scientists accelerate discovery, entrepreneurs launch new companies, and organizations operate with dramatically increased productivity.

This interaction between humans and intelligent machines may become one of the defining characteristics of the intelligence economy.

Key Takeaways

• Artificial intelligence could drive one of the largest economic expansions in modern history.

• AI technologies scale rapidly through digital networks, allowing innovations to spread quickly.

• Productivity gains across many industries may dramatically increase global economic output.

• Healthcare, energy, logistics, manufacturing, and scientific research may experience major transformation.

• The global competition to lead AI development is becoming a defining strategic challenge of the twenty-first century.

Expert Insight

Technological revolutions often unfold over decades, gradually reshaping industries, economic systems, and global power structures.

Artificial intelligence may represent the next stage of this historical pattern.

As machine intelligence becomes more capable, it may accelerate scientific discovery, increase economic productivity, and enable entirely new categories of technology.

The scale of this transformation may rival that of the industrial revolution or the rise of the internet.

Strategic Question

If artificial intelligence drives one of the largest economic expansions in history, who will benefit most from the new wealth it creates?

Will the intelligence economy empower entrepreneurs and innovators around the world, or will its benefits concentrate among a small number of powerful institutions?

Looking Ahead

AI is not only reshaping industries and economies; it is also raising deeper questions about the future of human civilization.

In the final chapters of this book, we explore how machine intelligence may influence global power, ethical responsibility, and the long-term relationship between humans and intelligent systems.

PART IV

INDUSTRIES BEING REWRITTEN

The Big Idea

Technological revolutions are often easiest to understand when we observe how they transform real industries.

The steam engine reshaped transportation and manufacturing. Electricity revolutionized factories and urban infrastructure. The internet transformed communication, commerce, and information access across the globe.

AI is beginning to produce a similar transformation.

Rather than influencing a single sector, AI is spreading across nearly every industry simultaneously. Machine learning systems can analyze vast datasets, recognize patterns within complex environments, and generate insights that were previously difficult or impossible to obtain.

As a result, industries built around information, research, logistics, and complex decision-making are beginning to change.

In medicine, artificial intelligence is accelerating the discovery of new drugs and assisting physicians in diagnosing disease. In finance, machine learning systems are analyzing markets, detecting fraud, and improving risk management. In robotics, intelligent software is enabling machines to interact with the physical world in increasingly sophisticated ways.

These developments suggest that artificial intelligence may become a foundational technology across the global economy.

Industries that successfully integrate AI into their operations may achieve dramatic improvements in productivity, efficiency, and innovation. Those that fail to adapt may struggle to compete in an environment where intelligent systems increasingly support research, design, and decision-making.

At the same time, the integration of artificial intelligence raises important questions.

How will industries evolve as machines become capable of performing increasingly complex tasks? What new opportunities will emerge as AI systems assist scientists, engineers, doctors, and entrepreneurs? And how will organizations restructure themselves around technologies capable of accelerating discovery and automation?

The chapters In this section explore several Industries where the Impact of artificial intelligence is already becoming visible.

By examining these transformations, we gain a clearer picture of how machine intelligence may reshape the global economy in the decades ahead.

AI is not only creating new technologies.

It is rewriting the industries that define modern civilization.

Why This Matters Now

When a powerful new technology emerges, its effects rarely remain confined to a single sector.

Instead, it begins reshaping entire industries.

AI is already transforming fields such as medicine, finance, robotics, and scientific research. Systems capable of analyzing vast datasets and identifying complex patterns are opening new possibilities for discovery and innovation.

Some of the most significant impacts of AI may occur in industries where the stakes are highest: healthcare, infrastructure, and global financial systems.

The chapters in this section explore how machine intelligence is beginning to rewrite the rules of several critical industries.

"AI is the new electricity." —Andrew Ng

Chapter 11

Artificial Intelligence in Medicine and Drug Discovery

For centuries, medicine advanced through slow experimentation. Artificial intelligence may accelerate discovery at a pace never before possible.

Recent breakthroughs illustrate the potential of artificial intelligence in medical research. In 2021, researchers introduced an AI system capable of predicting the three-dimensional structures of proteins with remarkable accuracy. Understanding protein structures is critical for developing new medicines, and this breakthrough dramatically accelerated biological research that previously required years of laboratory experimentation.

AI is also improving the detection of diseases such as cancer, diabetic eye disease, and neurological disorders through medical imaging. Machine learning models trained on large datasets of medical scans can identify subtle patterns that may be difficult for the human eye to detect.

While these technologies are still evolving, they illustrate how machine intelligence may significantly accelerate medical discovery in the decades ahead.

A Transformation in Healthcare

Among the many industries being reshaped by artificial intelligence, healthcare may experience some of the most profound changes.

Medicine has always been a field driven by knowledge and discovery. Physicians rely on years of education, clinical experience, and scientific research to diagnose illnesses and determine appropriate treatments.

But the volume of medical knowledge has grown enormously.

Thousands of new research papers are published every week. Medical imaging technologies generate vast quantities of diagnostic data. Genomic sequencing produces complex biological datasets containing millions of data points.

No single physician can realistically absorb and analyze all of this information.

Artificial intelligence offers a potential solution.

Machine learning systems are uniquely suited for identifying patterns within large datasets. They can analyze images, detect subtle statistical relationships, and process enormous quantities of information far more quickly than human researchers.

As a result, AI is beginning to play an increasingly important role in medical research, diagnosis, and patient care.

Diagnosing Disease with AI

One of the earliest applications of artificial intelligence in medicine has been medical imaging.

Doctors frequently rely on imaging technologies such as X-rays, CT scans, and MRI scans to diagnose disease. Interpreting these images requires specialized training and experience.

Machine learning systems can be trained to recognize patterns within medical images by analyzing large datasets of previously labeled examples.

For instance, an AI system might be trained on thousands of radiology scans that have already been reviewed by experts. Over time, the system learns to recognize visual features associated with specific medical conditions.

In some cases, AI systems have demonstrated accuracy comparable to human specialists when detecting certain types of disease.

Researchers have explored applications such as:

Detecting early signs of cancer in radiology images

Identifying abnormalities in cardiac scans

Diagnosing eye diseases through retinal imaging

Detecting neurological disorders through brain scans

These systems do not replace physicians, but they can serve as valuable diagnostic assistants.

By highlighting areas of concern in medical images, AI tools may help doctors identify conditions earlier and more accurately.

Accelerating Drug Discovery

Another area where artificial intelligence is making significant progress is pharmaceutical research.

Developing new medicines is an extremely complex and expensive process.

Researchers must identify biological targets associated with diseases, design molecules capable of interacting with those targets, test potential compounds in laboratory experiments, and conduct extensive clinical trials.

This process often takes more than a decade and costs billions of dollars.

Artificial intelligence has the potential to accelerate several stages of drug discovery.

Machine learning systems can analyze large datasets of chemical compounds and biological interactions, identifying patterns that might suggest promising drug candidates.

AI models can also simulate how molecules may interact with specific proteins or cellular structures.

By narrowing the range of possible candidates before laboratory testing begins, these systems may dramatically reduce the time required to develop new treatments.

Some pharmaceutical companies are already using AI to design entirely new molecular structures that could serve as potential drugs.

While these technologies are still evolving, they may significantly accelerate medical research in the future.

Personalized Medicine

Every human body is unique.

Two patients with the same disease may respond differently to the same treatment due to differences in genetics, lifestyle, and overall health.

Traditional medical treatments are often based on large population studies, which determine what works best for the average patient.

Artificial intelligence may help enable a more personalized approach to medicine.

By analyzing individual patient data, including genetic information, medical history, and lifestyle factors, AI systems may help doctors tailor treatments to the specific needs of each patient.

This concept, known as personalized medicine, could improve treatment effectiveness while reducing side effects.

For example, machine learning models may help predict how a particular patient will respond to a specific medication.

Doctors could then choose treatments that are most likely to be effective for that individual.

As genomic sequencing becomes more widely available, AI may play an increasingly important role in interpreting genetic data and identifying personalized treatment strategies.

Hospital Operations and Efficiency

AI is not limited to research and diagnosis.

It is also being applied to improve the efficiency of healthcare systems.

Hospitals are complex organizations that must manage large numbers of patients, staff members, medical equipment, and logistical processes.

AI systems can analyze operational data to identify ways of improving efficiency.

For example, machine learning models may help hospitals:

Predict patient admission rates

Optimize staff scheduling

Manage medical supply inventories

Improve emergency room triage processes

These improvements may reduce wait times for patients and allow healthcare providers to deliver care more efficiently.

In many countries, healthcare systems face increasing pressure due to aging populations and rising medical costs.

AI-driven operational improvements could help healthcare systems meet these challenges.

AI-Assisted Surgery

Robotic technologies are also beginning to play a role in surgical procedures.

Some surgical systems allow doctors to perform highly precise operations using robotic instruments controlled through specialized interfaces.

Artificial intelligence may further enhance these systems by assisting with tasks such as image guidance, surgical planning, and real-time analysis during procedures.

For example, AI systems could analyze medical imaging data to help surgeons visualize complex anatomical structures before an operation begins.

During surgery, machine learning algorithms might monitor vital signs and provide alerts if unusual patterns are detected.

These technologies are designed to support surgeons rather than replace them.

Human expertise and judgment remain essential in medical procedures.

However, AI-assisted tools may improve precision and reduce the risk of complications.

Ethical Considerations in Medical AI

While artificial intelligence offers many potential benefits for healthcare, it also raises important ethical questions.

Medical decisions often involve complex considerations that extend beyond purely technical analysis.

Issues such as patient privacy, informed consent, and data security must be carefully managed when developing AI systems for healthcare.

Another concern involves algorithmic bias.

If AI systems are trained on datasets that do not adequately represent diverse populations, their predictions may be less accurate for certain groups of patients.

Ensuring fairness and reliability in medical AI systems is therefore an important priority for researchers and healthcare providers.

Regulatory agencies around the world are also working to develop frameworks for evaluating the safety and effectiveness of AI-driven medical technologies.

These regulations aim to ensure that new systems are thoroughly tested before being used in clinical settings.

The Future of Medicine

Artificial intelligence will not replace doctors.

But it may change how medicine is practiced.

AI systems can analyze enormous datasets, identify patterns that might otherwise go unnoticed, and assist medical professionals in making more informed decisions.

Doctors, nurses, and researchers will remain central to the healthcare system.

Their expertise, empathy, and ethical judgment cannot be replicated by machines.

However, by combining human expertise with intelligent technologies, healthcare providers may be able to deliver better treatments, diagnose diseases earlier, and accelerate the pace of medical discovery.

In the coming decades, the integration of artificial intelligence into healthcare may help address some of the most challenging problems facing modern medicine.

And for patients around the world, these advances could lead to longer, healthier lives.

Key Takeaways

• AI is improving diagnostic accuracy in medical imaging.

• Machine learning systems may accelerate drug discovery and pharmaceutical research.

• AI tools enable personalized treatment strategies based on patient data.

• Hospitals are using AI to improve operational efficiency and patient management.

• Ethical oversight and regulatory frameworks remain essential for medical AI.

Expert Insight

Medicine has always been a field driven by data and discovery. As biological research generates increasingly complex datasets, artificial intelligence may become an indispensable tool for interpreting that information.

Machine learning systems can analyze patterns within medical images, genomic sequences, and clinical records that might otherwise remain hidden.

By assisting researchers and physicians in analyzing these datasets, AI technologies may help accelerate the development of new treatments and improve patient outcomes.

Strategic Question

If artificial intelligence can accelerate medical discovery and improve diagnostic accuracy, how should societies balance innovation with patient safety and ethical responsibility?

Who should oversee the deployment of AI systems that influence life-and-death medical decisions?

Looking Ahead

Healthcare is only one of many industries being transformed by artificial intelligence.

The next chapter explores how AI is reshaping global financial systems and markets.

Chapter 12

Artificial Intelligence and the Future of Finance

Financial markets run on information. When machines can analyze information faster than any human, the nature of finance begins to change.

The Intelligence Layer of Modern Finance

Few industries rely on data as heavily as finance.

Financial markets generate enormous volumes of information every second. Stock prices fluctuate, currencies shift in value, commodities respond to geopolitical events, and economic indicators continuously update the global financial picture.

For decades, financial institutions have relied on computers to process this information.

But artificial intelligence is introducing something new.

Instead of simply processing financial data, machines are beginning to analyze, predict, and respond to it in increasingly sophisticated ways.

Banks, hedge funds, insurance companies, and financial technology startups are rapidly adopting AI systems capable of detecting patterns within massive datasets. These systems can analyze market movements, identify fraud, automate trading strategies, and improve risk management.

In many ways, artificial intelligence is becoming the intelligence layer of modern finance.

The Evolution of Algorithmic Trading

Financial markets have long been influenced by technology.

During the late twentieth century, trading began shifting from physical trading floors to electronic platforms. Orders that were once executed through human brokers could now be placed through computerized systems.

This change enabled the development of algorithmic trading, where computer programs automatically execute trades based on predefined rules.

These systems could react to market changes far faster than human traders.

However, traditional algorithmic trading systems relied on relatively simple strategies.

They followed rules such as:

> Buying when a stock price crossed a specific threshold

> Selling when a market indicator reached a certain level

> Adjusting positions based on predetermined mathematical models

Artificial intelligence allows for far more complex strategies.

Machine learning models can analyze historical data, identify patterns in market behavior, and adapt their strategies over time.

Rather than following fixed rules, AI-driven trading systems can learn from experience.

This ability to continuously improve makes them powerful tools in the highly competitive world of financial markets.

Hedge Funds and AI-Driven Investment Strategies

Many hedge funds and investment firms have begun integrating artificial intelligence into their investment strategies.

These organizations collect enormous quantities of financial data, including:

> Stock market prices

Corporate earnings reports

Economic indicators

Global news events

Social media sentiment

Machine learning models can analyze these datasets to identify patterns that may signal potential investment opportunities.

For example, AI systems might detect correlations between macroeconomic indicators and sector performance, or identify early signals of market volatility.

Some investment firms even use natural language processing systems to analyze news articles and financial reports in real time.

These systems can scan thousands of documents within seconds, identifying information that may affect financial markets.

By combining this data with traditional financial analysis, hedge funds hope to gain an informational advantage in competitive markets.

However, AI-driven investing also introduces new risks.

Financial markets are complex systems influenced by human behavior, economic policy, and unpredictable events.

Machine learning models may detect patterns in historical data that do not necessarily predict future outcomes.

For this reason, human oversight remains essential in AI-driven financial strategies.

Fraud Detection and Financial Security

Another important application of artificial intelligence in finance involves fraud detection.

Financial fraud is a major global problem.

Credit card fraud, identity theft, and cybercrime cost financial institutions billions of dollars every year.

Traditional fraud detection systems often relied on fixed rules.

For example, a transaction might be flagged if it exceeded a certain amount or occurred in an unusual location.

However, fraud schemes have become increasingly sophisticated.

Artificial intelligence allows financial institutions to detect suspicious behavior more effectively.

Machine learning systems can analyze patterns across millions of transactions, identifying subtle anomalies that might indicate fraudulent activity.

For instance, an AI model might detect unusual spending patterns associated with a compromised credit card.

The system can then flag the transaction or temporarily freeze the account until further verification is completed.

These systems operate continuously, analyzing financial activity in real time.

As financial transactions increasingly move into digital environments, AI-powered security systems are becoming an essential component of modern banking infrastructure.

AI in Consumer Banking

AI is also transforming everyday banking services.

Many banks have begun integrating AI-driven chat systems capable of assisting customers with common requests.

These systems can answer questions about account balances, process simple transactions, and guide users through financial services.

In addition to customer support, AI tools are being used to provide personalized financial insights.

By analyzing spending patterns and account activity, machine learning systems can help customers understand their financial habits.

Some banking applications now offer features such as:

Automatic spending categorization

Savings recommendations

Budget planning tools

Fraud alerts and transaction monitoring

These services allow banks to provide more personalized financial guidance to customers.

Over time, AI may play an even greater role in personal financial management.

Risk Management in an Uncertain World

Risk management is one of the most important responsibilities of financial institutions.

Banks must carefully evaluate the risks associated with lending, investments, and market fluctuations.

AI is becoming an important tool in this process.

Machine learning models can analyze large datasets of financial information to estimate the probability of various economic outcomes.

For example, AI systems may help banks evaluate the creditworthiness of borrowers by analyzing factors such as income history, spending behavior, and economic conditions.

Insurance companies are also using machine learning models to estimate risk and price policies more accurately.

However, financial institutions must also be cautious when relying on AI models.

Because these systems learn from historical data, they may not always predict events that have never occurred before.

The global financial crisis of 2008 demonstrated how unexpected economic shocks can disrupt even the most sophisticated financial models.

As a result, financial institutions must balance technological innovation with careful oversight.

The Rise of Financial Technology

AI is also fueling the growth of financial technology, often referred to as fintech.

Fintech companies are startups focused on applying technology to financial services.

These companies often compete with traditional banks by offering faster, more accessible digital financial products.

Examples include:

Mobile payment platforms

Online lending services

Automated investment platforms

Digital banking applications

Artificial intelligence plays an important role in many of these services.

For instance, automated investment platforms use machine learning models to help users build diversified portfolios based on their financial goals and risk tolerance.

Online lending platforms use AI to evaluate loan applications quickly, sometimes approving loans within minutes.

These innovations are expanding access to financial services for millions of people around the world.

Financial Markets in the Age of AI

The growing influence of artificial intelligence is changing how financial markets operate.

Market participants now include not only human traders but also sophisticated algorithms capable of executing trades in milliseconds.

This technological evolution has increased market efficiency in many ways.

Prices adjust more quickly to new information, and trading costs have declined significantly over time.

However, the rise of algorithmic trading also raises important questions about market stability.

Highly automated markets may react extremely quickly to unexpected events, potentially amplifying volatility.

Regulators and financial institutions must therefore monitor how AI-driven trading systems interact with the broader financial ecosystem.

Ensuring that markets remain stable and fair is an important responsibility in an increasingly automated financial environment.

The Future of Finance

AI is unlikely to eliminate human involvement in financial markets.

Investment decisions often involve strategic thinking, risk tolerance, and judgment about future events that cannot be fully captured by mathematical models.

However, AI will almost certainly continue to influence how financial institutions operate.

Machines will analyze more data, detect patterns faster, and assist financial professionals in making more informed decisions.

The financial industry has always evolved alongside technological innovation.

From telegraphs and electronic trading systems to digital banking and global payment networks, technology has continually reshaped financial markets.

Artificial intelligence represents the next stage in that evolution.

And as AI systems become more sophisticated, they may help create a financial system that is faster, more efficient, and more responsive to the needs of a rapidly changing global economy.

Key Takeaways

• AI systems analyze financial markets and detect patterns in large datasets.

• Algorithmic trading strategies increasingly rely on machine learning.

• Financial institutions use AI to detect fraud and manage risk.

• Fintech startups are expanding access to digital financial services.

• Artificial intelligence may help create faster and more efficient financial systems.

Expert Insight

Financial markets operate on information. The ability to analyze and interpret that information quickly can provide significant competitive advantages.

Artificial intelligence systems excel at analyzing large volumes of structured and unstructured data. They can process financial reports, news articles, and market indicators far faster than human analysts.

This capability allows financial institutions to identify emerging trends and respond to market changes in real time.

However, AI systems must be used carefully to avoid amplifying financial instability or reinforcing existing biases in economic models.

Strategic Question

If artificial intelligence can analyze financial markets faster and more accurately than human traders, how will this influence the stability of global financial systems?

Could AI-driven markets become more efficient, or more volatile?

Looking Ahead

Beyond finance, artificial intelligence is also entering the physical world.

The next chapter explores the rise of robotics and physical AI systems.

Chapter 13

Robotics and the Physical AI Revolution

Artificial intelligence does not have to remain inside computers. When combined with robotics, it begins to reshape the physical world.

Intelligence Leaves the Screen

For much of its modern history, artificial intelligence has existed primarily in the digital world.

AI systems analyzed data, generated text, recognized images, and assisted with decision-making inside computers and software platforms. Their influence was powerful, but largely confined to digital environments.

Now that is beginning to change.

AI is moving beyond software and entering the physical world.

When AI systems are combined with machines capable of interacting with the environment, including robots, autonomous vehicles, and intelligent manufacturing equipment, the result is a new technological frontier sometimes referred to as physical AI.

This shift may transform industries ranging from manufacturing and logistics to agriculture, transportation, and infrastructure.

The AI revolution is no longer limited to algorithms.

It is becoming mechanical.

The Long Evolution of Robotics

Robotics has existed for decades, particularly in manufacturing environments.

Industrial robots first appeared in factories during the 1960s. These machines were designed to perform repetitive tasks such as welding, painting, and assembling components on production lines.

Unlike modern AI-driven robots, these early systems were not truly intelligent.

They followed highly precise instructions programmed by engineers. Every movement was predetermined, and even small changes in the environment could disrupt the robot's operation.

For example, if a part on an assembly line was slightly misaligned, the robot might fail to complete its task correctly.

Despite these limitations, industrial robots dramatically increased manufacturing productivity.

Automobile factories, electronics assembly plants, and heavy machinery facilities all benefited from robotic automation.

However, these machines were limited to highly controlled environments.

The next generation of robotics aims to overcome that limitation.

AI Gives Robots Perception

One of the biggest challenges in robotics has always been perception.

Humans possess remarkable sensory capabilities. We can recognize objects, interpret complex environments, and adapt to unexpected situations.

Early robots lacked these abilities.

AI is changing that.

Machine learning systems trained on large datasets of images and sensor data can now help robots interpret their surroundings.

Computer vision systems allow robots to identify objects, detect obstacles, and track movement.

Advanced sensors such as lidar, radar, and depth cameras provide additional environmental information.

By combining these technologies, robots can build detailed representations of their surroundings and make decisions based on that information.

This ability to perceive the environment is a crucial step toward creating machines that can operate outside of controlled factory settings.

Robotics in Modern Manufacturing

Manufacturing remains one of the industries most heavily influenced by robotics.

Modern factories increasingly rely on advanced robotic systems capable of performing tasks with extraordinary precision.

These machines can assemble complex products, handle delicate components, and operate continuously without fatigue.

AI is enhancing these capabilities.

AI-driven robots can adapt to variations in materials or components, adjusting their actions in real time rather than relying on rigid programming.

For example, machine learning systems may allow robots to identify defects in products during manufacturing, improving quality control.

Predictive maintenance systems can analyze equipment data to identify potential mechanical problems before they cause production delays.

As these technologies continue to evolve, manufacturing facilities may become increasingly autonomous.

Some factories already operate with minimal human intervention, using networks of robotic systems to manage production processes.

Warehouse Automation

One of the most visible applications of robotics and AI can be found in modern warehouses.

The growth of e-commerce has dramatically increased the demand for efficient logistics systems.

Companies must store, locate, package, and ship millions of products quickly and accurately.

Robotic systems are increasingly used to perform these tasks.

Autonomous warehouse robots can move through storage facilities, retrieving items from shelves and transporting them to packing stations.

Machine learning systems help these robots navigate complex environments, avoid collisions, and coordinate their movements with other robots.

Advanced inventory management systems track the location of every item within the warehouse.

These technologies allow logistics companies to process orders far more efficiently than traditional manual systems.

As e-commerce continues to grow, robotic warehouse automation is likely to expand further.

Autonomous Vehicles

Perhaps the most widely discussed application of physical AI involves autonomous vehicles.

Self-driving cars have long been a goal of robotics researchers and automotive engineers.

Driving requires a combination of perception, decision-making, and motor control.

Vehicles must interpret road conditions, recognize traffic signals, anticipate the behavior of other drivers, and respond to unexpected situations.

Artificial intelligence systems trained on vast datasets of driving scenarios are beginning to demonstrate these capabilities.

Autonomous vehicle systems typically combine several technologies:

Computer vision systems that analyze camera images

Lidar sensors that create detailed maps of surroundings

Machine learning models that predict the behavior of nearby vehicles and pedestrians

Navigation systems that plan optimal driving routes

Although fully autonomous driving remains a complex challenge, progress in this field has been significant.

Some cities are already testing autonomous taxis and delivery vehicles.

If these technologies become widely adopted, they could reshape transportation systems around the world.

Robotics in Agriculture

Another area where robotics and AI are gaining traction is agriculture.

Modern farms face increasing pressure to produce food efficiently while managing limited land and environmental resources.

Robotic systems may help address these challenges.

Agricultural robots can perform tasks such as:

Planting crops

Monitoring soil conditions

Identifying weeds and pests

Harvesting produce

Machine learning systems analyze sensor data to determine the optimal timing for irrigation, fertilization, and harvesting.

Precision agriculture techniques allow farmers to apply resources only where they are needed, reducing waste and improving crop yields.

As global populations continue to grow, these technologies may play an important role in ensuring sustainable food production.

The Rise of Humanoid Robots

One of the most ambitious areas of robotics research involves humanoid robots.

These machines are designed with physical structures resembling the human body, allowing them to navigate environments designed for human workers.

Humanoid robots could potentially perform a wide variety of tasks in industries such as:

Manufacturing

Construction

Healthcare

Disaster response

Developing these systems presents significant engineering challenges.

Balancing on two legs, manipulating objects with dexterous hands, and adapting to unpredictable environments require sophisticated control systems and advanced AI.

However, progress in machine learning and robotics engineering is gradually bringing these capabilities closer to reality.

Some companies are already testing humanoid robots capable of performing simple tasks in warehouses and industrial environments.

While widespread adoption may still be years away, the development of humanoid robotics represents an important frontier in the AI revolution.

Safety and Human Collaboration

As robots become more capable, ensuring safety becomes increasingly important.

Early industrial robots were typically isolated behind protective barriers to prevent accidents.

Modern collaborative robots, sometimes called cobots, are designed to operate safely alongside human workers.

These systems use sensors and AI algorithms to detect human presence and adjust their movements accordingly.

For example, a robot working on an assembly line might slow down or stop entirely if a person approaches too closely.

Collaborative robots allow humans and machines to work together in shared environments, combining the strengths of both.

Robots can handle repetitive or physically demanding tasks, while humans provide judgment, flexibility, and problem-solving abilities.

The Future of Physical AI

The integration of artificial intelligence and robotics represents one of the most exciting frontiers of modern technology.

As machine learning systems become more capable and hardware continues to improve, robots may gradually expand into new industries and applications.

Physical AI could transform sectors such as:

Construction and infrastructure development

Healthcare and elder care

Environmental monitoring

Space exploration

While the timeline for these developments remains uncertain, the direction is clear.

AI is no longer confined to digital systems.

It is beginning to take physical form.

And as machines become capable of interacting with the world around them, the AI revolution may reshape not only the digital economy but the physical world as well.

Key Takeaways

• AI is enabling robots to perceive and interact with their environments.

• Robotics systems are transforming manufacturing, logistics, and agriculture.

• Autonomous vehicles represent one of the most ambitious applications of physical AI.

• Humanoid robots may eventually operate in human-designed environments.

• Human-robot collaboration may become increasingly common in many industries.

Expert Insight

Combining artificial intelligence with robotics represents one of the most significant technological frontiers of the modern era.

While software-based AI systems operate in digital environments, robots bring machine intelligence into the physical world. This integration allows machines to perform tasks involving movement, perception, and interaction with complex environments.

Advances in computer vision, sensor technology, and machine learning have dramatically improved robotic capabilities. Over time, these systems may expand into industries ranging from construction to healthcare.

Strategic Question

As robotics and artificial intelligence merge to create machines capable of operating in the physical world, how will societies integrate intelligent systems into environments traditionally dominated by human labor?

Which industries will experience the earliest and most profound transformation?

Looking Ahead

As AI systems grow more capable, researchers are beginning to ask an even more ambitious question.

Could machines eventually achieve general intelligence?

The next chapter explores the concept of Artificial General Intelligence.

PART V

THE FRONTIER OF INTELLIGENCE

The Big Idea

AI is advancing rapidly, but the systems used today may represent only the earliest stages of machine intelligence.

The algorithms currently transforming industries are powerful, yet they remain specialized. Most AI systems excel at specific tasks such as generating language, analyzing images, or predicting patterns within large datasets.

The frontier of artificial intelligence lies beyond these narrow capabilities.

Researchers are increasingly exploring the possibility of Artificial General Intelligence (AGI), systems capable of performing a wide range of intellectual tasks at levels comparable to or exceeding human abilities.

If such systems are eventually developed, they could dramatically accelerate scientific discovery, technological innovation, and economic growth.

At the same time, the emergence of more advanced forms of machine intelligence raises profound questions.

How should societies govern technologies that may become extraordinarily powerful? What safeguards are necessary to ensure that AI systems operate safely and responsibly? And how might the development of advanced AI influence the balance of global power among nations?

These questions are no longer purely theoretical.

Governments, research institutions, and technology companies are already debating how to manage the development of increasingly capable AI systems. Policies surrounding regulation, safety research, and international cooperation may shape the trajectory of artificial intelligence for decades to come.

The frontier of Intelligence therefore represents both an opportunity and a responsibility.

Artificial intelligence could help solve some of humanity's most complex challenges, from disease and climate modeling to scientific discovery and advanced engineering.

Yet the same technologies could also introduce risks if developed or deployed without careful oversight.

The chapters in this section explore the forces shaping the future of machine intelligence.

They examine the technological foundations that power advanced AI systems, the path toward more general forms of artificial intelligence, the ethical considerations surrounding these technologies, and the geopolitical dynamics that may emerge as nations compete for leadership in the AI era.

AI is not just transforming the present.

It is shaping the frontier of the future.

And the choices made during the early stages of this transformation may determine how the age of intelligent machines ultimately unfolds.

Why This Matters Now

Artificial intelligence has already begun transforming industries and economic systems.

But the most profound developments may still lie ahead.

Researchers around the world are exploring the next generation of intelligent systems, technologies capable of more advanced reasoning, autonomous problem solving, and deeper collaboration with humans. These developments may lead toward systems that approach broader forms of intelligence.

The pursuit of these technologies is not merely a scientific challenge.

It is also a strategic one.

Governments, technology companies, and research institutions are investing enormous resources to develop increasingly capable AI systems. The outcomes of this competition may influence economic power, national security, and global technological leadership for decades to come.

At the same time, the rapid advancement of machine intelligence raises important ethical and societal questions.

How should powerful AI systems be governed?

What safeguards should exist to ensure their responsible use?

And how might the development of increasingly capable machines influence the balance between human decision-making and automated systems?

The chapters in this section explore these questions.

They examine the technological frontier of artificial intelligence, the global competition to develop advanced systems, and the responsibilities that accompany the creation of machines capable of influencing the future of human civilization.

"The development of full artificial intelligence could spell the end of the human race... or the beginning of something extraordinary."

—Stephen Hawking

Chapter 14

The Power Behind Artificial Intelligence

Behind every intelligent system lies immense computational power.

The Hidden Architecture of Intelligence

When people interact with artificial intelligence, the experience often feels simple.

A user types a question.

An AI system produces an answer.

To the person using the technology, the interaction appears almost effortless.

But behind that seemingly simple exchange lies one of the most complex technological infrastructures ever built.

Modern artificial intelligence systems depend on enormous networks of computing hardware, vast collections of data, sophisticated algorithms, and global digital platforms capable of delivering intelligent services to billions of users.

These layers form what some researchers describe as the AI power stack.

Understanding this stack is essential to understanding how artificial intelligence will shape the future.

Because the organizations that control these layers may ultimately influence how machine intelligence develops, and who benefits from it.

Artificial intelligence may be one of the most transformative technologies in human history.

But it is also becoming one of the most strategically important.

The First Layer: The Chip Revolution

At the foundation of the AI power stack lies hardware.

Artificial intelligence requires enormous computational power. Training modern machine learning systems involves billions, sometimes trillions, of mathematical operations.

These calculations are performed by specialized processors designed to handle large-scale parallel computation.

For many years, graphics processing units, or GPUs, became the dominant hardware platform for artificial intelligence.

Originally developed for rendering video game graphics, GPUs turned out to be remarkably effective at performing the matrix operations required by neural networks.

As machine learning research advanced, GPUs became the engines powering modern AI development.

Today, large AI models are often trained using thousands of GPUs operating together in massive computing clusters.

These clusters perform calculations at speeds that would have been unimaginable just a few decades ago.

Because of this, access to advanced semiconductor technology has become a critical factor in the global AI race.

Countries and corporations now recognize that control over advanced chips may influence who leads the next generation of technological innovation.

Artificial intelligence begins not with algorithms, but with silicon.

The Second Layer: Infrastructure

If chips are the engines of artificial intelligence, infrastructure is the power grid that keeps those engines running.

Training and operating large AI models requires vast computing facilities known as data centers.

Inside these facilities are thousands, sometimes tens of thousands, of servers connected through high-speed networks.

These machines process enormous volumes of data and perform the calculations required to train machine learning models.

Some of the largest AI training clusters in the world consume as much electricity as small cities.

Cooling systems circulate massive quantities of air and liquid through server racks to prevent processors from overheating.

This infrastructure is expensive to build and operate.

Constructing a single advanced data center can cost hundreds of millions of dollars.

As artificial intelligence becomes more powerful, these facilities are expanding rapidly across the world.

The companies capable of building and maintaining these networks may gain enormous advantages in the AI economy.

In many ways, the development of artificial intelligence resembles earlier technological revolutions.

Just as the industrial revolution required railroads, factories, and electrical grids, the AI revolution requires data centers, cloud computing networks, and global digital infrastructure.

The Third Layer: The Rise of Foundation Models

Above the hardware and infrastructure layers lies the software architecture that makes artificial intelligence possible.

Over the past decade, researchers have begun developing what are known as foundation models.

These are large machine learning systems trained on enormous datasets containing text, images, code, and other forms of digital information.

Foundation models are capable of performing many different tasks using a single underlying architecture.

They can generate written text, answer questions, summarize documents, translate languages, analyze images, and assist with software development.

What makes these systems particularly powerful is their flexibility.

Unlike earlier AI systems that were designed for specific tasks, foundation models can adapt to many different applications.

This versatility has made them central to the current wave of artificial intelligence innovation.

However, building these models requires enormous resources.

Training a cutting-edge AI model may require months of computation across thousands of processors.

The cost of training such systems can reach tens or even hundreds of millions of dollars.

Because of this, only a small number of organizations currently possess the resources required to develop the largest AI models.

This concentration of capability raises important questions about the future structure of the AI ecosystem.

The Fourth Layer: Data

Artificial intelligence systems learn by analyzing data.

The quality, diversity, and scale of the data used to train AI models play a critical role in determining how well those systems perform.

Over the past two decades, the internet has created an unprecedented volume of digital information.

Text documents, images, videos, scientific research papers, and software code now exist in vast quantities online.

Machine learning systems can analyze these datasets to learn patterns and relationships.

In many cases, the performance of an AI model improves as the amount of training data increases.

However, data also introduces new challenges.

Issues of privacy, copyright, and ethical data usage are becoming increasingly important as AI systems expand.

Researchers are also exploring new approaches such as synthetic data, where artificial intelligence generates its own training datasets to supplement existing information.

The future of AI development may depend not only on computing power, but also on how societies manage access to data.

The Fifth Layer: Distribution

The final layer of the AI power stack is distribution.

Even the most advanced artificial intelligence system has little impact unless it can reach users.

Distribution platforms determine how AI tools are integrated into everyday life.

These platforms may include:

Productivity software used by businesses

Cloud computing services used by developers

Mobile applications used by consumers

Enterprise platforms used by large organizations

When artificial intelligence becomes embedded within widely used software systems, its influence expands rapidly.

Millions, or even billions, of users can interact with AI tools through applications they already use daily.

This distribution layer is where much of the economic value of artificial intelligence may ultimately be captured.

Companies that control the platforms through which AI services are delivered may shape how the technology is adopted across society.

The Strategic Importance of the AI Stack

When viewed together, these layers reveal something important.

AI is not simply a software innovation.

It is an entire technological ecosystem composed of hardware, infrastructure, models, data, and distribution platforms.

Each layer contributes to the development of machine intelligence.

And each layer represents a potential source of strategic advantage.

Nations that invest in semiconductor manufacturing may strengthen their position in the hardware layer.

Technology companies that build massive cloud computing networks may dominate the infrastructure layer.

Organizations that develop powerful AI models may influence the direction of future innovation.

Meanwhile, companies that control major software platforms may determine how billions of people interact with artificial intelligence.

This interconnected structure makes the AI revolution far more complex than many earlier technological developments.

AI is not owned by a single entity.

But its power is distributed across a network of organizations competing to build the systems that will define the future.

The Emerging AI Power Structure

As artificial intelligence continues to evolve, the organizations capable of controlling key parts of the AI stack may gain enormous influence.

The competition to build these systems is already intensifying.

Technology companies are investing billions of dollars into research, infrastructure, and advanced computing facilities.

Governments are developing national strategies to support domestic AI development.

Universities and research laboratories continue pushing the boundaries of machine learning and computational science.

The result is a rapidly expanding global ecosystem focused on one objective:

Building the next generation of intelligent systems.

A New Strategic Resource

Throughout history, societies have competed for control over critical resources.

Land, energy, and industrial production once determined the balance of power between nations.

In the twenty-first century, intelligence itself may become one of the most valuable resources.

Artificial intelligence systems have the potential to accelerate scientific discovery, increase economic productivity, and transform entire industries.

The organizations capable of building and deploying these systems may shape the trajectory of technological progress for decades to come.

Understanding the AI power stack therefore reveals something deeper.

The development of artificial intelligence is not simply a story about machines becoming smarter.

It is also a story about who builds those machines, and who controls the systems that power them.

And in the coming decades, that question may become one of the most important strategic questions in the world.

Key Takeaways

• Artificial intelligence systems depend on massive computational resources.

• Specialized hardware such as GPUs and AI accelerators enable large-scale machine learning.

• Semiconductor technology has become a strategic asset in the global AI race.

• Data centers and high-performance computing clusters form the backbone of modern AI development.

• Access to computing infrastructure increasingly determines which organizations can build the most advanced AI systems.

Expert Insight

Behind every breakthrough in artificial intelligence lies an enormous technological infrastructure. Training modern machine learning models requires vast computing clusters, specialized processors, and sophisticated networking systems capable of coordinating billions of calculations simultaneously.

This infrastructure has transformed artificial intelligence from a purely academic discipline into an industrial-scale engineering effort. Companies developing advanced AI models now invest billions of dollars into computing hardware and data center capacity.

As AI systems grow larger and more capable, the availability of computing resources may become one of the defining factors that determines which organizations lead the next wave of technological innovation.

Strategic Question

If computing power is the foundation of modern artificial intelligence, how will the race to build more powerful processors shape the future of technology?

Will breakthroughs in computing architecture unlock entirely new levels of machine intelligence?

Looking Ahead

While computing power provides the foundation for artificial intelligence, the ultimate goal of many researchers extends far beyond current machine learning systems.

The next chapter explores one of the most ambitious ideas in the field: the development of Artificial General Intelligence, or AGI, systems capable of performing a broad range of intellectual tasks at a level comparable to human reasoning.

Understanding the path toward AGI is essential for understanding the long-term trajectory of artificial intelligence.

Chapter 15

The Path Toward Artificial General Intelligence

If machines continue improving their reasoning abilities, they may eventually match or surpass human intelligence.

Beyond Specialized Machines

Most artificial intelligence systems today are highly specialized.

They are designed to perform specific tasks such as recognizing images, translating languages, recommending products, or generating text. These systems can be extremely powerful within their designated domains, but they lack the broader adaptability that characterizes human intelligence.

A chess-playing AI, for example, may defeat the best human players in the world, yet it cannot drive a car, write a business strategy, or diagnose a medical condition without being retrained for those tasks.

Human intelligence, by contrast, is general.

People can learn new skills, apply knowledge across disciplines, and adapt to unfamiliar situations.

The idea of creating machines with similar versatility has long fascinated scientists and philosophers.

This concept is known as Artificial General Intelligence, often abbreviated as AGI.

AGI refers to AI systems capable of performing a wide range of intellectual tasks at or above human levels of competence.

Such systems would not simply execute predefined algorithms. They would be capable of reasoning, learning, and adapting across many domains.

If achieved, AGI would represent one of the most profound technological breakthroughs in human history.

The Difference Between Narrow AI and AGI

To understand the significance of AGI, it is helpful to distinguish it from the systems that exist today.

Most modern AI applications fall into the category of narrow AI, also called weak AI.

Narrow AI systems are highly effective at specific tasks but cannot easily transfer their capabilities to unrelated areas.

For example:

A language model can generate text but cannot operate a robot without additional systems.

A computer vision system can identify objects in images but cannot analyze financial markets.

A recommendation algorithm can suggest movies but cannot design a medical treatment.

These systems rely on specialized training datasets and architectures optimized for particular applications.

AGI, by contrast, would possess broader cognitive flexibility.

An AGI system might be able to:

Learn new skills with minimal training

Solve problems across different fields

Adapt to unfamiliar environments

Combine knowledge from multiple disciplines

In essence, AGI would resemble a general-purpose intellect rather than a specialized tool.

Why AGI Is So Difficult

Despite decades of research, achieving AGI remains an extremely difficult challenge.

Human intelligence arises from a complex biological system consisting of approximately 86 billion neurons connected through trillions of synapses.

The brain integrates perception, memory, reasoning, emotion, and motor control into a unified cognitive system.

Modern AI models, while powerful, still operate in fundamentally different ways.

Large neural networks can recognize patterns in enormous datasets, but they do not possess the same depth of understanding or contextual awareness that humans exhibit.

For example, language models can generate convincing explanations of scientific concepts, yet they may struggle with basic reasoning problems that humans find intuitive.

This gap highlights the difference between statistical pattern recognition and true general reasoning.

Bridging this gap remains one of the central challenges of AI research.

The Role of Scaling

One of the most significant discoveries in recent AI research involves what are known as scaling laws.

Researchers observed that increasing the size of neural networks, the amount of training data, and the computational resources used during training often leads to improved performance.

Larger models tend to exhibit more sophisticated capabilities.

This observation has led to the development of extremely large AI models trained on vast datasets using massive computing clusters.

Some researchers believe that continuing to scale these models could eventually lead to systems with general intelligence.

According to this view, AGI may emerge gradually as models grow more powerful and capable.

Others argue that scaling alone will not be sufficient.

They believe that entirely new architectures or cognitive frameworks may be required to achieve true general intelligence.

The debate continues among AI researchers today.

Competing Paths to AGI

Several different approaches are being explored in the search for artificial general intelligence.

One approach focuses on improving existing deep learning models by increasing their size and training them on more diverse datasets.

Another approach involves combining neural networks with symbolic reasoning systems capable of performing logical operations.

Some researchers are exploring biologically inspired models that more closely resemble the structure of the human brain.

Others are investigating reinforcement learning systems capable of learning through interaction with complex environments.

Each of these approaches offers potential advantages.

However, none has yet produced a system that fully matches the versatility of human intelligence.

The path to AGI may ultimately involve a combination of these techniques.

Timelines and Predictions

One of the most debated questions in artificial intelligence is when AGI might be achieved.

Predictions vary widely.

Some researchers believe AGI could emerge within the next few decades.

Others argue that it may take much longer, possibly a century or more.

Part of the difficulty lies in the unpredictable nature of scientific breakthroughs.

Major discoveries can occur suddenly after years of incremental progress.

In the early 20th century, few people anticipated how rapidly aviation and space technology would advance.

Similarly, the pace of AI progress has accelerated dramatically in recent years.

While it is impossible to predict exactly when AGI will arrive, it is clear that the field is moving quickly.

Governments, corporations, and research institutions are investing enormous resources into AI development.

The incentives for achieving major breakthroughs are substantial.

Potential Benefits of AGI

If artificial general intelligence were successfully developed, the potential benefits could be extraordinary.

AGI systems could assist researchers in solving some of humanity's most complex challenges.

They might accelerate scientific discovery, design new medical treatments, optimize energy systems, and develop innovative technologies.

For example, AGI could potentially help:

Discover cures for currently incurable diseases

Develop new materials with revolutionary properties

Improve climate modeling and environmental management

Optimize transportation and infrastructure systems

In many ways, AGI could function as a powerful intellectual partner for human researchers.

By processing vast amounts of information and generating new insights, such systems might dramatically accelerate the pace of human progress.

The Risks of Advanced Intelligence

However, the development of AGI also raises important questions about safety and governance.

A machine intelligence capable of performing complex reasoning could potentially influence many aspects of society.

Ensuring that such systems operate safely and align with human values is therefore a major focus of ongoing research.

AI safety researchers study how advanced systems can be designed to behave reliably, even in complex or unpredictable environments.

This work involves questions such as:

How can AI systems interpret human intentions accurately?

How can they avoid unintended harmful outcomes?

What safeguards should exist for extremely powerful technologies?

Addressing these challenges will likely require collaboration among scientists, engineers, policymakers, and ethicists.

The Threshold of a New Intelligence

The pursuit of artificial general intelligence represents one of the most ambitious scientific goals of our time.

Humanity is attempting to create a new form of intelligence, one that exists not in biological organisms but in machines.

Whether AGI emerges in decades or centuries, the research efforts underway today are laying the groundwork for that possibility.

Each improvement in machine learning models, computing infrastructure, and data analysis brings researchers closer to understanding how intelligent systems function.

If AGI is eventually achieved, it may mark a turning point in human history.

For the first time, intelligence would no longer be limited to biological life.

It would exist in machines as well.

And the consequences of that development could reshape civilization in ways we are only beginning to imagine.

Key Takeaways

• Artificial General Intelligence (AGI) refers to systems capable of performing a wide range of intellectual tasks.

• Most current AI systems remain specialized rather than general.

• Achieving AGI would represent one of the most significant breakthroughs in technological history.

• Researchers continue debating the timeline and feasibility of AGI development.

• AGI raises profound questions about the future relationship between humans and machines.

Expert Insight

The concept of AGI has fascinated scientists and philosophers for decades. While modern machine learning systems demonstrate impressive capabilities in specific domains, replicating the full flexibility of human intelligence remains an enormous challenge.

Human cognition integrates perception, reasoning, memory, creativity, and emotional understanding within a single system. Building machines with similar versatility requires advances in many areas of artificial intelligence research.

Although AGI may still be years or decades away, the rapid progress of AI technologies suggests that continued breakthroughs are possible.

Strategic Question

If artificial general intelligence eventually becomes possible, how should humanity define the boundaries of machine autonomy?

What responsibilities come with creating systems capable of reasoning at or beyond human levels?

Looking Ahead

If machines eventually achieve broader forms of intelligence, society will face complex ethical questions.

The next chapter examines the ethical challenges of artificial intelligence.

Chapter 16

The Ethics of Artificial Intelligence

The rise of intelligent machines raises questions humanity has never had to confront before.

Technology and Responsibility

Throughout history, powerful technologies have always raised ethical questions.

The invention of electricity transformed industry but required new safety standards. The development of nuclear energy created immense power but also introduced risks that demanded international oversight. The rise of the internet connected billions of people but also created new challenges involving privacy, security, and misinformation.

Artificial intelligence may represent a similar turning point.

As AI systems become more capable and integrated into everyday life, they increasingly influence important decisions about finance, healthcare, employment, and public policy.

This growing influence raises a critical question:

> How should intelligent machines be designed, governed, and used responsibly?

The answer to this question may shape the future relationship between humans and artificial intelligence.

The Challenge of Algorithmic Bias

One of the most widely discussed ethical issues in artificial intelligence involves algorithmic bias.

Machine learning systems learn patterns from data.

If the data used to train these systems contains biases or imbalances, the resulting AI models may reflect those biases in their predictions.

For example, an AI system trained on historical hiring data may learn patterns that reflect past discrimination or unequal representation in the workforce.

Similarly, facial recognition systems trained on limited datasets may perform less accurately when analyzing individuals from underrepresented demographic groups.

These issues do not arise because machines possess intentions or beliefs.

They arise because machine learning systems mirror the patterns present in the data used to train them.

Addressing algorithmic bias therefore requires careful attention to how datasets are constructed, evaluated, and updated.

Researchers are actively working to develop techniques that improve fairness and reduce bias in AI systems.

However, ensuring fairness remains a complex challenge.

Transparency and Explainability

Another ethical concern involves the transparency of AI systems.

Many modern machine learning models operate as complex neural networks with millions or billions of parameters.

These models can produce highly accurate predictions, but understanding exactly how they arrive at those predictions can be difficult.

This lack of transparency is sometimes referred to as the "black box" problem.

For example, if an AI system recommends denying a loan application or predicts that a patient may have a serious medical condition, decision-makers may want to understand the reasoning behind that prediction.

Researchers are therefore developing techniques known as explainable AI.

These methods attempt to provide insights into how AI systems reach their conclusions.

Explainable AI may help build trust in machine learning technologies by allowing users to better understand the factors influencing algorithmic decisions.

However, achieving full transparency in highly complex models remains an ongoing research challenge.

Privacy in the Age of Data

Artificial intelligence systems often rely on enormous datasets for training and operation.

These datasets may include personal information such as online behavior, purchasing patterns, location data, and medical records.

While this data can enable powerful AI capabilities, it also raises concerns about privacy.

Individuals may not always be aware of how their data is being collected or used.

Companies and organizations must therefore implement safeguards to protect personal information and ensure that data is used responsibly.

Many governments have introduced regulations aimed at protecting data privacy.

These regulations often require organizations to obtain user consent before collecting personal data and to provide transparency about how that data is used.

Balancing innovation with privacy protection will remain an important challenge as AI technologies continue to evolve.

The Responsibility of Developers

Artificial intelligence systems do not exist independently.

They are designed, trained, and deployed by human organizations.

As a result, developers and companies bear significant responsibility for ensuring that their technologies are used safely and ethically.

Responsible AI development often involves several key principles:

Careful evaluation of training data

Rigorous testing of system behavior

Monitoring systems after deployment

Implementing safeguards against misuse

Organizations developing AI technologies must also consider the broader societal impact of their systems.

A tool that improves efficiency in one context may create unintended consequences in another.

For example, automated decision-making systems used in hiring or financial services could influence economic opportunities for individuals.

Developers must therefore carefully consider how their technologies affect the communities that use them.

Regulation and Governance

As artificial intelligence becomes more influential, governments around the world are beginning to develop regulatory frameworks for AI technologies.

These frameworks aim to address concerns such as safety, accountability, and fairness.

Regulation can serve several purposes.

It can establish standards for testing AI systems before they are deployed in critical environments. It can require transparency in automated decision-making processes. It can also provide mechanisms for addressing harm if AI systems produce unintended outcomes.

However, regulating AI presents unique challenges.

Technology evolves rapidly, while legal systems often move more slowly.

Regulators must therefore design policies that protect public interests without unnecessarily slowing innovation.

Achieving this balance will likely require ongoing collaboration between policymakers, researchers, and technology companies.

AI Safety Research

Another area of growing importance is AI safety research.

This field focuses on ensuring that advanced AI systems behave reliably and align with human goals.

Researchers in this area explore questions such as:

How can AI systems interpret human instructions accurately?

How can they avoid unintended harmful actions?

What safeguards should exist for extremely powerful AI technologies?

While many current AI systems operate within relatively narrow domains, future systems may become significantly more capable.

Ensuring that these systems remain safe and beneficial is therefore a major focus of ongoing research.

Organizations across academia and industry are investing in efforts to better understand how intelligent systems behave and how their behavior can be guided responsibly.

The Role of Society

The ethical development of artificial intelligence cannot be left solely to engineers and policymakers.

AI technologies affect society as a whole.

Public discussions about AI ethics involve questions about fairness, accountability, and the values that should guide technological progress.

Educators, researchers, business leaders, and citizens all play a role in shaping how AI systems are used.

Public understanding of artificial intelligence is therefore important.

The more people understand how these technologies work, the better society can participate in decisions about their development and deployment.

A Shared Responsibility

AI is one of the most powerful technologies humanity has ever created.

It has the potential to transform industries, accelerate scientific discovery, and improve quality of life across the world.

But like all powerful technologies, it must be developed with care.

Ensuring that AI systems are safe, fair, and beneficial will require cooperation across many disciplines.

Computer scientists must design reliable algorithms. Policymakers must develop thoughtful regulations. Businesses must adopt responsible practices. And society must remain engaged in conversations about how these technologies should shape the future.

The ethical challenges of artificial intelligence are not obstacles to progress.

They are an essential part of guiding that progress in a direction that benefits humanity.

Key Takeaways

• Artificial intelligence raises important ethical and societal questions.

• Concerns include bias, privacy, transparency, and accountability.

• Responsible AI development requires collaboration between researchers, governments, and industry.

• Ethical frameworks are being developed to guide AI deployment.

• Public trust will be essential for the successful integration of AI technologies.

Expert Insight

Ethical considerations are central to the future of artificial intelligence. As AI systems influence decisions related to healthcare, finance, employment, and governance, ensuring fairness and accountability becomes increasingly important.

Developers must carefully examine how AI models are trained and deployed to prevent unintended consequences such as algorithmic bias or misuse of personal data.

Establishing ethical standards for AI development may help ensure that technological progress benefits society as a whole.

Strategic Question

As artificial intelligence becomes more powerful, how can societies ensure that these technologies remain aligned with human values?

Who should determine the ethical principles that guide the development of intelligent machines?

Looking Ahead

Beyond ethics, artificial intelligence also carries significant geopolitical implications.

The next chapter explores how AI may influence global power and international relations.

Chapter 17

Artificial Intelligence and Global Power

Technology has always shaped geopolitical power. Artificial intelligence may become the most important strategic technology of the 21st century.

The Strategic Resource of the Twenty-First Century

For most of human history, the balance of global power has been determined by physical resources.

In ancient civilizations, fertile land and agricultural productivity defined the strength of nations. Empires rose and fell based on their ability to control territory capable of sustaining large populations.

During the industrial revolution, the strategic resources shifted. Coal, steel, and later oil became the engines of national power. Countries that controlled industrial production gained enormous economic and military advantages.

In the late twentieth century, information networks and digital technology began reshaping global influence. Nations capable of developing advanced computing infrastructure and communication networks became the leaders of the emerging digital economy.

Now, a new strategic resource is emerging.

Intelligence itself.

Artificial intelligence systems have the potential to transform how societies generate knowledge, make decisions, and deploy technology. As these systems become more capable, they may influence nearly every sector of modern civilization, from scientific research and economic productivity to military strategy and geopolitical competition.

For the first time at global scale, intelligence is becoming a scalable resource.

And the nations that control it may shape the future of global power.

Crossing the Intelligence Threshold

Human civilization has long operated under a simple assumption.

Intelligence belongs to biological organisms.

Every major technological achievement, from the construction of ancient cities to the exploration of space, has ultimately depended on human reasoning.

But artificial intelligence is beginning to challenge that assumption.

When machines can analyze enormous datasets, generate new ideas, assist with scientific discovery, and coordinate complex systems, intelligence itself becomes a form of infrastructure.

Humanity may now be approaching what could be described as the Intelligence Threshold.

The Intelligence Threshold represents a moment in history when intelligence is no longer limited to biological minds. Instead, it becomes embedded within machines, networks, and global computing systems.

Once this threshold is crossed, the production of knowledge may accelerate dramatically.

Scientific discoveries could occur faster. Economic decisions could be optimized through advanced data analysis. Military systems could coordinate complex operations using intelligent software.

In effect, intelligence becomes a technological capability rather than a purely biological trait.

This shift could redefine how power operates in the modern world.

The AI Superpowers

Today, several regions are competing to lead the development of artificial intelligence.

The United States currently possesses one of the strongest AI ecosystems. American universities produce leading research in machine learning, and many of the most influential technology companies are headquartered within the United States.

These organizations control enormous computing infrastructure, advanced semiconductor design capabilities, and large pools of technical talent.

China represents another major force in artificial intelligence development. Over the past decade, Chinese technology companies and government institutions have invested heavily in AI research, robotics, and advanced computing infrastructure.

China's vast digital economy generates massive datasets, providing valuable training material for machine learning systems.

Europe has taken a somewhat different approach, emphasizing ethical governance and regulatory frameworks for artificial intelligence. European institutions have focused on ensuring that AI technologies are developed in ways that protect privacy, fairness, and human rights.

While these regions differ in strategy, they share a common recognition:

Artificial intelligence may become one of the most strategically important technologies of the twenty-first century.

Infrastructure as Power

Behind the visible advances in artificial intelligence lies an enormous physical infrastructure.

Training modern AI systems requires massive computing clusters composed of thousands of specialized processors operating simultaneously.

These clusters consume enormous amounts of electricity and require advanced cooling systems to operate efficiently.

Technology companies are now constructing some of the largest computing facilities ever built.

In many ways, these facilities resemble the industrial factories of earlier technological revolutions.

But instead of producing steel or automobiles, they produce something far more abstract.

They produce intelligence.

Access to large-scale computing infrastructure has therefore become one of the most important factors in the global AI race.

The organizations capable of training the most powerful models may gain significant advantages in innovation, economic productivity, and scientific discovery.

Control over the infrastructure of intelligence may become as important as control over energy resources was in earlier eras.

The Military Implications of AI

AI is also beginning to influence military strategy.

Modern defense systems increasingly rely on advanced software capable of processing vast amounts of information in real time.

Machine learning systems can analyze satellite imagery, monitor cybersecurity threats, and assist with decision-making in complex operational environments.

Autonomous technologies are also becoming more sophisticated.

Robotic systems capable of operating in air, sea, and land environments are already being developed. Some of these systems may eventually operate with increasing levels of autonomy.

These developments raise significant ethical and strategic questions.

How should societies govern autonomous weapons?

What safeguards should exist to prevent unintended escalation in conflicts involving intelligent systems?

The answers to these questions will play an important role in shaping the future of global security.

Economic Power in the AI Era

Beyond military applications, artificial intelligence may also reshape the structure of the global economy.

AI systems have the potential to dramatically increase productivity by automating complex analytical tasks and accelerating research and development.

Industries that adopt these technologies effectively may gain significant competitive advantages.

Countries that lead in AI innovation may therefore experience stronger economic growth, increased technological leadership, and greater influence over emerging industries.

This dynamic could produce new forms of economic competition.

Just as the industrial revolution reshaped global manufacturing, the AI revolution may reshape the global knowledge economy.

The most advanced economies of the future may not be those with the largest factories.

They may be those with the most advanced intelligence infrastructure.

The Risks of an Intelligence Divide

While artificial intelligence offers enormous opportunities, it also raises concerns about inequality.

If advanced AI capabilities are concentrated within a small number of countries or corporations, a significant technological gap could emerge.

Nations without access to advanced computing infrastructure or skilled researchers may struggle to compete in an increasingly AI-driven world.

This could lead to what some analysts describe as an intelligence divide.

Countries capable of deploying advanced AI systems may accelerate their economic and scientific development, while others fall further behind.

Managing this divide may become one of the major geopolitical challenges of the coming decades.

International cooperation, knowledge sharing, and responsible governance may play important roles in ensuring that the benefits of artificial intelligence are broadly distributed.

The Future Balance of Power

Predicting the long-term impact of artificial intelligence on global power is difficult.

Technological revolutions rarely follow simple trajectories. Unexpected breakthroughs, new political alliances, and emerging industries can reshape the landscape in unpredictable ways.

However, one trend appears increasingly clear.

AI is becoming a foundational technology.

Like electricity, the internet, and the industrial machinery of earlier eras, AI may eventually influence nearly every aspect of modern society.

The nations and organizations that lead in its development may gain extraordinary influence over the direction of technological progress.

The global balance of power in the twenty-first century may therefore depend not only on economic resources or military strength.

It may depend on who controls the systems that generate intelligence itself.

And that competition is only beginning.

Key Takeaways

• AI development is becoming a major factor in geopolitical competition.

• Nations view AI leadership as essential for economic and strategic power.

• Control of computing infrastructure and semiconductor technology is increasingly important.

• AI capabilities may influence military, economic, and technological leadership.

• Global cooperation and competition will shape the AI landscape.

Expert Insight

AI is not only a technological transformation but also a geopolitical one. Nations that successfully develop advanced AI infrastructure, talent pipelines, and research ecosystems may gain long-term strategic advantages. The global AI race is therefore not simply about innovation It is about influence over the future technological order.

Strategic Question

If artificial intelligence becomes a decisive factor in national power, how should countries balance technological competition with international cooperation?

Could AI become the defining geopolitical force of the twenty-first century?

Looking Ahead

The geopolitical competition surrounding artificial intelligence will shape global economic systems as well.

In the next chapter, we explore how AI may transform the global economy and redefine productivity, industry, and wealth creation in the decades ahead.

PART VI

THE WORLD AHEAD

The Big Idea

Technological revolutions do more than introduce new tools.

They reshape the structure of society.

The printing press transformed how knowledge spread across civilizations. The industrial revolution altered how economies produced goods and organized labor. The digital revolution connected billions of people through global networks of information.

Artificial intelligence may trigger a transformation of similar magnitude.

But unlike earlier technologies that primarily amplified physical labor or information exchange, AI expands something even more fundamental.

It expands intelligence itself.

As machine learning systems become more capable, they may assist with scientific discovery, accelerate engineering innovation, improve healthcare diagnostics, optimize complex infrastructure systems, and help solve challenges that previously required decades of human research.

These developments could reshape the global economy, redefine the nature of work, and influence how societies make decisions about technology, governance, and innovation.

Yet the future of artificial intelligence is not predetermined.

The trajectory of this technology will be shaped by the choices made by researchers, entrepreneurs, policymakers, and citizens around the world.

Will artificial intelligence be developed in ways that expand opportunity and accelerate progress?

Or will its benefits become concentrated among a small number of institutions and organizations?

Will humans learn to collaborate effectively with intelligent machines?

Or will societies struggle to adapt to rapid technological change?

The answers to these questions will influence the direction of the intelligence revolution.

The chapters in this final section explore what the coming decades may hold.

They examine the future economic systems that could emerge in an AI-driven world, the evolving relationship between humans and intelligent machines, and the long-term possibilities that may unfold as machine intelligence continues to advance.

The goal is not to predict the future with certainty.

Rather, it is to explore the possibilities that lie ahead, and to consider the role each of us may play in shaping them.

The intelligence revolution is still in its early chapters.

The world that emerges from it will depend on what we choose to build next.

Why This Matters Now

Every technological revolution forces societies to reconsider how the future may unfold.

AI is no exception.

The technologies emerging today may transform the structure of the global economy, reshape how humans collaborate with machines, and influence the trajectory of scientific discovery.

Yet the future of artificial intelligence is not predetermined.

The decisions made by governments, researchers, entrepreneurs, and citizens will shape how this technology evolves.

The chapters in this final section explore possible futures for a world increasingly influenced by machine intelligence.

"We are entering a world where intelligence becomes a tool we can build."

—*Fei-Fei Li*

Chapter 18

The Future Economy in an AI World

Some of the most valuable companies in the world are now competing to build the intelligence infrastructure of the future.

A New Economic Engine

Technological revolutions often reshape the global economy.

The steam engine powered the industrial revolution, transforming agriculture and manufacturing. Electricity enabled mass production and modern infrastructure. The internet connected global markets and created entirely new industries.

Artificial intelligence may become the next great economic engine.

By automating complex tasks, analyzing massive datasets, and assisting with decision-making, AI has the potential to dramatically increase productivity across many sectors of the economy.

Productivity growth is one of the most important drivers of economic progress.

When workers are able to produce more value in less time, societies can generate higher incomes, create new industries, and improve living standards.

Artificial intelligence may therefore play a central role in shaping the economic landscape of the twenty-first century.

Productivity and the Knowledge Economy

Modern economies are increasingly driven by knowledge-based work.

Many professions involve analyzing information, solving complex problems, and developing new ideas.

Examples include:

Engineering

Scientific research

Finance and investment

Software development

Design and creative industries

Business strategy and management

Artificial intelligence systems can assist with many of these tasks by processing information at extraordinary speed.

Machine learning models can analyze financial markets, simulate engineering designs, and identify patterns in scientific data that might otherwise take years to uncover.

By accelerating these processes, AI may allow professionals to accomplish more within the same amount of time.

A scientist working with AI tools may analyze far larger datasets than previously possible.

An entrepreneur may use AI systems to design prototypes and analyze market opportunities quickly.

In this way, artificial intelligence may function as a powerful productivity multiplier.

The Emergence of AI-Driven Industries

As artificial intelligence becomes more capable, it may also give rise to entirely new industries.

Some of these industries are already beginning to emerge.

Companies are developing AI-powered software tools for business operations, healthcare diagnostics, autonomous transportation, and scientific research.

New markets are forming around AI infrastructure, including specialized hardware, cloud computing platforms, and data management systems.

Other industries may arise as AI capabilities continue to evolve.

For example, intelligent personal assistants could become central components of everyday life, helping individuals manage schedules, finances, education, and health information.

AI-driven research tools may accelerate discoveries in fields such as biotechnology, materials science, and climate science.

Over time, these developments may reshape the global economic structure.

Just as the internet gave rise to digital media, e-commerce, and social networking industries, artificial intelligence may produce new economic sectors that do not yet exist.

The Distribution of Economic Benefits

While technological innovation often increases overall economic productivity, it can also create challenges related to the distribution of benefits.

When new technologies automate certain types of work, workers in those sectors may face economic disruption.

For example, automation in manufacturing during the twentieth century significantly increased productivity but also reduced the number of workers required for certain industrial jobs.

Artificial intelligence may produce similar effects in knowledge-based professions.

Some tasks previously performed by humans may become partially automated.

This could lead to shifts in employment patterns across industries.

However, technological revolutions also create new forms of work.

As AI systems expand into new applications, they may generate demand for roles involving system design, data analysis, AI management, and human-machine collaboration.

The challenge for societies will be ensuring that workers have opportunities to develop the skills required for emerging industries.

Education and workforce training will therefore play a crucial role in the transition to an AI-driven economy.

The Debate Over Universal Basic Income

As discussions about AI-driven automation continue, some economists and policymakers have proposed new economic policies designed to address potential workforce disruptions.

One widely discussed concept is Universal Basic Income (UBI).

Under a UBI system, citizens would receive a regular financial payment from the government regardless of employment status.

Supporters argue that such policies could provide financial stability in a future where automation reduces the availability of certain types of work.

Critics, however, raise concerns about the economic feasibility and potential social consequences of universal income programs.

The debate over UBI reflects broader questions about how societies should adapt to technological change.

While artificial intelligence may create enormous economic value, policymakers must consider how that value is distributed and how workers are supported during periods of transition.

Global Competition for Economic Leadership

AI is also becoming an important factor in global economic competition.

Countries investing heavily in AI research, infrastructure, and education may gain advantages in productivity and technological innovation.

These advantages could influence international trade, industrial development, and economic growth.

Governments around the world are therefore implementing strategies aimed at strengthening their AI ecosystems.

These strategies often include investments in research laboratories, technology startups, semiconductor manufacturing, and advanced education programs.

In the coming decades, leadership in artificial intelligence may become one of the defining characteristics of economically powerful nations.

Small Businesses and AI

Artificial intelligence will not only affect large corporations.

Small businesses and independent entrepreneurs may also benefit from AI-powered tools.

For example, small companies can use AI systems to analyze customer behavior, optimize marketing strategies, and automate routine administrative tasks.

Online platforms increasingly provide AI-driven services that help businesses manage logistics, financial planning, and customer communication.

These tools allow small organizations to operate more efficiently, even with limited resources.

In some cases, a small team equipped with advanced AI tools may compete effectively with much larger organizations.

This democratization of technology may encourage innovation and entrepreneurship.

The Acceleration of Innovation

Another important economic effect of artificial intelligence may involve the acceleration of scientific and technological innovation.

Researchers often spend large amounts of time analyzing data, conducting simulations, and testing hypotheses.

AI systems can assist with many of these tasks.

Machine learning models can analyze experimental data, identify patterns, and generate new research hypotheses.

In fields such as chemistry and materials science, AI systems can simulate molecular interactions and predict the properties of new compounds.

These capabilities may significantly accelerate the pace of scientific discovery.

New technologies that might have taken decades to develop could potentially emerge much more quickly.

This acceleration of innovation may further amplify the economic impact of artificial intelligence.

The Structure of the AI Economy

The emerging AI economy may consist of several interconnected layers.

At the foundation lies infrastructure, including semiconductor manufacturing, data centers, and cloud computing networks.

Above that layer are AI platforms, which provide machine learning models and tools that developers can use to build applications.

At the top are AI-powered applications, which serve specific industries such as healthcare, finance, education, and logistics.

This layered structure resembles the architecture of the internet economy.

Infrastructure companies provide the technical foundation, platform providers offer development tools, and application developers create services for users.

Understanding this structure may help entrepreneurs and investors identify opportunities within the AI ecosystem.

A Transformative Economic Shift

Artificial intelligence may become one of the most transformative economic forces of the twenty-first century.

By increasing productivity, accelerating innovation, and enabling new forms of entrepreneurship, AI could significantly expand the global economy.

At the same time, societies will need to navigate the challenges associated with technological change.

Workforces must adapt to new skill requirements. Governments must consider policies that support economic transitions. Businesses must learn how to integrate intelligent technologies responsibly.

The economic future shaped by artificial intelligence is still unfolding.

But the foundations of that future are already being built.

And the decisions made today may influence the prosperity of societies for generations to come.

Key Takeaways

• Artificial intelligence may dramatically increase economic productivity.

• Entire industries may evolve as AI becomes integrated into business operations.

• New markets and business models are likely to emerge.

• Economic transitions may require adaptation from workers and institutions.

• AI could become a central driver of global economic growth.

Expert Insight

Technological revolutions often trigger new waves of economic expansion. By automating routine tasks and accelerating research, artificial intelligence may unlock new forms of productivity across many sectors.

Entrepreneurs, researchers, and businesses will likely explore innovative ways to integrate AI into products and services.

However, managing the transition toward an AI-driven economy will require thoughtful policy decisions and investment in education and workforce development.

Strategic Question

If artificial intelligence dramatically increases productivity across industries, how will societies distribute the economic benefits of that growth?

Will AI create greater prosperity for many, or greater concentration of wealth for a few?

Looking Ahead

As AI becomes integrated into the economy, humans will increasingly interact with intelligent machines.

The next chapter explores the future of human-AI collaboration.

Chapter 19

Human AI Collaboration

Investors are pouring billions into artificial intelligence. They believe the technology could reshape the global economy.

Intelligence as a Partnership

When new technologies emerge, they are often viewed as replacements for human labor.

During the industrial revolution, machines replaced manual production in factories. Agricultural equipment replaced many traditional farming tasks. More recently, computers automated many administrative and analytical functions.

Artificial intelligence has sparked similar concerns.

Many people worry that intelligent machines could eventually replace human workers in a wide range of professions.

However, the most powerful applications of artificial intelligence may not involve replacing humans at all.

Instead, the future may be defined by collaboration between human intelligence and machine intelligence.

Humans and AI systems possess different strengths.

When these strengths are combined effectively, the result can be far more powerful than either working alone.

The Strengths of Machines

Artificial intelligence systems excel at tasks involving large volumes of data and repetitive computation.

They can analyze massive datasets within seconds, identify statistical patterns that might escape human attention, and perform calculations with extraordinary speed.

For example, AI systems can:

Process millions of financial transactions to detect fraud

Analyze complex engineering simulations

Scan medical images for subtle diagnostic patterns

Search vast databases of scientific research

These capabilities allow machines to handle tasks that would be time-consuming or impossible for humans to perform manually.

However, machines also have limitations.

Despite their computational power, AI systems lack many of the qualities associated with human reasoning and judgment.

The Strengths of Humans

Human intelligence remains uniquely suited for certain kinds of tasks.

People possess creativity, intuition, ethical reasoning, and emotional understanding.

These qualities are essential in areas such as leadership, strategic planning, education, and artistic expression.

Humans also excel at navigating complex social environments.

Negotiation, persuasion, and interpersonal communication require subtle understanding of human behavior and context.

While AI systems can assist with many analytical tasks, they do not possess the lived experiences, cultural awareness, and moral judgment that shape human decision-making.

The combination of human insight and machine computation therefore creates powerful opportunities for collaboration.

Augmented Intelligence

Rather than viewing artificial intelligence as a replacement for human expertise, many researchers now prefer the term augmented intelligence.

Augmented intelligence refers to technologies designed to enhance human capabilities rather than replace them.

In this model, AI systems serve as tools that amplify human creativity, productivity, and analytical power.

For example, an architect might use AI-assisted design software to explore hundreds of structural variations for a building project.

A scientist might use machine learning models to analyze experimental data and identify promising research directions.

A business strategist might use AI tools to simulate market scenarios and evaluate potential outcomes.

In each case, the AI system performs large-scale analysis, while the human expert interprets the results and makes final decisions.

This collaborative approach allows both human and machine strengths to complement each other.

AI in Scientific Research

One of the most promising areas for human-AI collaboration is scientific research.

Modern scientific problems often involve extremely complex datasets.

Fields such as genomics, climate science, and particle physics generate enormous volumes of information that must be analyzed carefully.

Artificial intelligence can assist researchers by identifying patterns within these datasets and suggesting potential hypotheses.

For example, machine learning models can analyze genetic sequences to identify potential disease-related mutations.

In materials science, AI systems can simulate chemical interactions and suggest new compounds with useful properties.

These capabilities allow researchers to explore possibilities much more quickly than traditional methods would allow.

However, human scientists remain essential for interpreting results, designing experiments, and evaluating the broader implications of discoveries.

The collaboration between AI systems and human researchers may accelerate the pace of scientific progress.

Creative Collaboration

AI is also beginning to influence creative industries.

Artists, musicians, writers, and designers are experimenting with AI tools capable of generating images, music, and text.

These tools do not replace human creativity.

Instead, they can serve as creative partners.

A designer might use AI systems to generate visual concepts and then refine them into a final artwork.

A musician might use AI tools to explore new melodies or sound patterns.

Writers can use AI systems to brainstorm ideas or generate draft outlines for stories and essays.

In this way, artificial intelligence becomes a tool for expanding creative possibilities.

Rather than limiting artistic expression, it may provide new forms of inspiration.

Decision-Making in Complex Systems

Another important area of human-AI collaboration involves complex decision-making.

Large organizations often face decisions involving enormous amounts of data and many uncertain variables.

For example, urban planners may need to evaluate transportation systems, housing development, environmental impact, and economic growth when designing city infrastructure.

Artificial intelligence can assist by modeling different scenarios and analyzing the potential outcomes of various policy choices.

Human decision-makers can then use this information to guide strategic planning.

Similarly, businesses may use AI systems to analyze supply chains, predict market trends, and optimize resource allocation.

In these situations, AI functions as an analytical engine, while human leaders provide judgment and direction.

Education and Learning

Human-AI collaboration may also transform education.

Traditional education systems often rely on standardized instruction delivered to large groups of students.

Artificial intelligence could enable more personalized learning experiences.

AI-driven educational platforms can analyze student performance data and adapt lessons to match individual learning styles and pace.

For example, a student struggling with a particular mathematical concept might receive additional exercises and explanations tailored to their needs.

Meanwhile, students who progress quickly may move ahead to more advanced topics.

Teachers remain central to the educational process.

Their role may shift toward mentoring, guiding discussions, and supporting students' intellectual development.

AI tools can assist with grading, lesson planning, and personalized feedback.

Together, teachers and AI systems may create more effective learning environments.

The Human Role in an Intelligent World

As artificial intelligence becomes more integrated into society, the human role may evolve rather than disappear.

People may increasingly focus on activities that require creativity, empathy, leadership, and ethical judgment.

AI systems may handle many routine analytical tasks, freeing humans to concentrate on higher-level thinking and collaboration.

This shift could lead to new forms of work that combine human creativity with machine intelligence.

Rather than competing with AI systems, humans may learn to partner with them.

The result may be a new form of productivity where intelligent machines extend the reach of human potential.

A Collaborative Future

The relationship between humans and artificial intelligence is still developing.

Early fears about automation often focus on the possibility that machines will replace human labor entirely.

However, history suggests that the most transformative technologies tend to reshape human work rather than eliminate it.

Artificial intelligence may follow this pattern.

By combining computational power with human creativity and judgment, AI could become a powerful partner in solving complex problems.

Scientists, engineers, artists, entrepreneurs, and educators may all find new ways to collaborate with intelligent systems.

Together, humans and machines may accomplish goals that neither could achieve alone.

And this partnership may become one of the defining features of the AI-powered world.

Key Takeaways

• Human-AI collaboration may become a central feature of future work environments.

• AI systems can assist with research, analysis, and creative tasks.

• Humans remain essential for leadership, ethics, and complex decision-making.

• Effective collaboration between humans and machines may increase innovation.

• The future of work may involve partnerships between people and intelligent systems.

Expert Insight

The relationship between humans and machines is evolving. Rather than viewing AI solely as a replacement for human labor, many researchers envision a future in which intelligent systems augment human capabilities.

By assisting with data analysis, design, and problem solving, AI tools may allow individuals to focus on higher-level creative and strategic work.

This collaborative model could unlock new forms of productivity and innovation.

Strategic Question

If humans and intelligent machines begin working together as collaborative partners, how might creativity, research, and problem-solving evolve?

Could human-AI collaboration unlock forms of innovation that neither could achieve alone?

Looking Ahead

As artificial intelligence becomes integrated into many aspects of society, it is worth imagining what the world might look like in the decades ahead.

The next chapter explores a possible vision of life in 2040.

Chapter 20

The 2040 Intelligence Economy

If current trends continue, the global economy of 2040 could look radically different from today.

Imagining the Next Two Decades

Predicting the future has always been difficult.

Technological progress rarely unfolds in perfectly predictable ways. Breakthroughs often arrive unexpectedly, while other innovations take far longer to develop than anticipated.

Yet certain technological trends provide clues about how the coming decades might evolve.

AI is advancing rapidly. Machine learning models are becoming more capable, computing infrastructure is expanding, and intelligent systems are increasingly integrated into everyday tools and services.

By examining these trends, it is possible to imagine what the world might look like in the year 2040 if artificial intelligence continues its current trajectory.

While such projections are necessarily speculative, they offer insight into the potential scale of the transformation underway.

Ten Transformations That Artificial Intelligence May Bring

By 2040, artificial intelligence may reshape many aspects of human civilization. While the exact timeline remains uncertain, several technological trajectories suggest how machine intelligence could transform society in the coming decades.

AI-Assisted Scientific Discovery

AI systems may help researchers discover new materials, medicines, and energy technologies far faster than traditional methods.

Personal AI Assistants

Individuals may rely on highly capable digital assistants capable of managing schedules, conducting research, and coordinating complex tasks.

Autonomous Transportation Networks

Self-driving vehicles could dramatically reduce traffic accidents and transform logistics systems.

AI-Driven Healthcare

Medical AI systems may assist doctors in diagnosing diseases earlier and recommending personalized treatment plans.

One-Person Global Companies

Entrepreneurs equipped with AI tools may build companies that once required large teams of employees.

AI-Designed Products and Engineering Systems

Artificial intelligence could design complex machinery, architecture, and infrastructure systems.

Intelligent Energy Systems

AI may help optimize power grids and accelerate the development of renewable energy technologies.

Human-AI Collaborative Workplaces

Rather than replacing humans, many AI systems will likely function as collaborators assisting with research, analysis, and decision-making.

Accelerated Education

AI tutors may provide personalized education for students around the world.

New Ethical and Governance Challenges

As AI systems become more powerful, societies will face complex decisions about regulation, responsibility, and the ethical development of intelligent technologies.

The world of 2040 will almost certainly look different from the world today.

But the trajectory is already visible.

AI is beginning to reshape how knowledge is created, how businesses operate, and how societies solve complex problems.

And the story of this transformation is only just beginning.

Intelligent Cities

Cities may be among the first environments to experience large-scale integration of artificial intelligence.

Urban areas generate enormous amounts of data through transportation networks, energy systems, communication infrastructure, and public services.

AI-driven systems could help manage these complex networks more efficiently.

For example, intelligent traffic systems might analyze real-time traffic data to optimize traffic signals and reduce congestion.

Autonomous public transportation systems could coordinate buses, trains, and shared vehicles to improve mobility across metropolitan areas.

Energy grids may use machine learning models to balance electricity demand and supply more efficiently, particularly as renewable energy sources such as solar and wind become more widespread.

Smart infrastructure could monitor the condition of bridges, roads, and buildings, identifying maintenance needs before serious problems occur.

These technologies may help cities become more efficient, sustainable, and resilient.

Healthcare in the AI Era

Healthcare may also undergo dramatic changes as artificial intelligence continues to advance.

Medical AI systems could assist physicians in diagnosing diseases earlier and more accurately by analyzing medical images, genetic information, and patient health data.

Personalized medicine may become more common as machine learning systems analyze individual genetic profiles to recommend treatments tailored to each patient.

Hospitals may rely on AI-driven logistics systems to manage patient flow, optimize staffing schedules, and coordinate the use of medical equipment.

Remote healthcare services may expand through the use of AI-powered diagnostic tools and telemedicine platforms.

These technologies could make healthcare more accessible, particularly in regions where medical professionals are scarce.

While human doctors and nurses will remain essential, AI systems may help them provide more effective and efficient care.

The Evolution of Work

By 2040, the nature of work may look quite different from today.

Artificial intelligence systems will likely continue to automate many routine tasks, both physical and cognitive.

However, new forms of work may also emerge.

Many professions may involve collaboration between humans and intelligent systems.

Engineers might rely on AI tools to simulate complex designs and optimize performance. Scientists may work alongside machine learning systems that analyze massive datasets and suggest new research directions.

Entrepreneurs may use AI-driven platforms to develop new products, manage business operations, and reach global markets more easily.

Education systems may also evolve to prepare individuals for this new environment.

Skills such as creativity, critical thinking, interdisciplinary knowledge, and collaboration with intelligent tools may become increasingly important.

Rather than eliminating human work, artificial intelligence may transform the types of tasks that humans perform.

Autonomous Systems Everywhere

Autonomous systems may become more common in many aspects of daily life.

Transportation networks may include self-driving cars, autonomous delivery vehicles, and AI-coordinated logistics systems.

Warehouses, factories, and agricultural operations may rely heavily on robotic systems capable of performing tasks with minimal human supervision.

Drones could assist with infrastructure inspection, environmental monitoring, and emergency response.

These technologies may increase efficiency in industries such as transportation, construction, agriculture, and logistics.

However, they will also require careful regulation and oversight to ensure safety and reliability.

The transition toward widespread autonomy is likely to occur gradually, as technologies are tested, refined, and integrated into existing systems.

The Acceleration of Scientific Discovery

One of the most exciting possibilities for artificial intelligence lies in its potential to accelerate scientific discovery.

Many scientific breakthroughs require analyzing enormous datasets and exploring complex interactions between variables.

AI systems are particularly well suited for this type of analysis.

Machine learning models can process vast amounts of information and identify patterns that may lead to new insights.

In fields such as medicine, materials science, and environmental research, AI-driven tools may help scientists develop new technologies more quickly.

For example, AI systems might assist in designing advanced battery materials, discovering new pharmaceutical compounds, or modeling climate systems.

If these capabilities continue to improve, the pace of innovation across many scientific disciplines could accelerate dramatically.

Everyday Life with AI Assistants

Artificial intelligence may also become a routine part of everyday life.

Personal AI assistants could help individuals manage schedules, finances, education, and health information.

These systems might coordinate travel plans, recommend learning resources, and monitor health data collected from wearable devices.

For example, an AI assistant might remind a user to schedule a medical checkup based on health data patterns or suggest educational courses aligned with career goals.

These tools could help individuals make more informed decisions and manage complex aspects of modern life.

However, ensuring privacy and data security will be essential as such systems become more integrated into personal routines.

Challenges Ahead

Despite its potential benefits, the future shaped by artificial intelligence will also present challenges.

Societies must address questions related to employment transitions, economic inequality, data privacy, and the ethical use of advanced technologies.

Governments, businesses, and educational institutions will need to work together to ensure that technological progress benefits as many people as possible.

Public discussions about the responsible use of AI will remain important as these technologies become more influential.

The goal should not be to slow innovation, but to guide it in ways that promote human well-being and long-term stability.

The Shape of Tomorrow

The world of 2040 will likely look different from the world of today.

Artificial intelligence may become deeply embedded in infrastructure, healthcare, transportation, research, and everyday digital tools.

At the same time, human creativity, leadership, and ethical judgment will remain central to shaping how these technologies are used.

The future of AI is not predetermined.

It will be shaped by the choices made by researchers, entrepreneurs, policymakers, and citizens around the world.

The technologies we build today may influence how societies function for generations to come.

And the story of artificial intelligence is still only beginning.

The Future We Choose

Every generation experiences moments when the course of history begins to shift.

For those living through such transitions, the change is not always immediately obvious. Progress often arrives quietly at first, as an experiment in a laboratory, a prototype in a research facility, an unexpected breakthrough that gradually spreads across industries and societies.

Artificial intelligence appears to be one of those moments.

What began as a field of theoretical research has evolved into a transformative technological force. Machines capable of analyzing information, generating ideas, and assisting with complex decision-making are now integrated into the daily lives of millions of people.

Yet we are still at the early stages of this transformation.

The systems that exist today represent only the first generation of modern artificial intelligence. Their capabilities continue to expand as researchers improve algorithms, develop more powerful computing infrastructure, and explore new ways for machines to learn from data.

The coming decades may bring developments that are difficult to imagine today.

Artificial intelligence could accelerate scientific discovery, enabling breakthroughs in medicine, energy, and materials science. It could improve global infrastructure, optimize transportation networks, and help address complex environmental challenges.

At the same time, the rise of intelligent machines introduces new responsibilities.

Technology does not shape the future on its own. The impact of artificial intelligence will depend on the choices made by the people who design, deploy, and govern it.

Researchers must ensure that intelligent systems are developed safely and responsibly. Businesses must consider the societal consequences of the technologies they create. Policymakers must craft regulations that encourage innovation while protecting the public interest.

And individuals must learn to adapt to a world where intelligence is no longer limited to human minds.

The story of artificial intelligence is therefore not only about machines.

It is about humanity.

It is about how we choose to use our knowledge, how we guide technological progress, and how we balance innovation with responsibility.

The age of machine intelligence has begun, but its final shape has not yet been determined.

The future of artificial intelligence is still being written.

And ultimately, it will be written by us.

Key Takeaways

• Artificial intelligence may dramatically reshape the global economy over the coming decades.

• Advances in automation, machine learning, and robotics could increase productivity across many industries.

• Economic systems may shift toward knowledge-driven and AI-assisted production.

• Nations that adapt their education, infrastructure, and policy frameworks may gain economic advantages.

• The long-term economic impact of AI will depend on how societies manage technological transitions.

Expert Insight

Technological revolutions rarely transform economies overnight, but they often alter the underlying structure of productivity. Artificial intelligence may represent one of the most powerful productivity technologies since the invention of electricity and the internet.

AI systems can accelerate research, optimize logistics networks, analyze complex datasets, and assist with decision-making across industries. These capabilities could significantly expand economic output.

However, the benefits of technological progress are rarely distributed evenly. Governments, businesses, and educational institutions will play a crucial role in ensuring that the opportunities created by AI are broadly accessible.

The societies that adapt successfully to this technological shift may define the economic landscape of the twenty-first century.

Strategic Question

If artificial intelligence dramatically increases productivity across industries, how should societies distribute the economic benefits of that growth?

Will AI create broader prosperity, or lead to new forms of economic concentration?

Looking Ahead

The economic transformation driven by artificial intelligence is only one part of a much larger story.

Beyond markets and productivity lies a deeper question about the relationship between humans and intelligent machines.

In the next chapter, we explore how human creativity, judgment, and collaboration may evolve in a world where artificial intelligence becomes a permanent partner in human progress.

Chapter 21

Designing the Intelligence Economy

The intelligence economy will not emerge randomly. It will be shaped by policy, innovation, and global competition.

A Glimpse Into the Next Technological Era

Predicting the future of technology is always uncertain.

Yet history shows that when powerful technologies reach a tipping point, their influence spreads rapidly across society. Electricity reshaped cities, factories, and transportation systems. The internet transformed communication, commerce, and access to information.

Artificial intelligence may represent the next great transformation.

If current technological trends continue, the world of 2040 may look fundamentally different from the one we inhabit today. Intelligent systems may assist with scientific discovery, manage complex infrastructure, and collaborate with humans across nearly every profession.

This emerging system can be described as the intelligence economy, a global network of human and machine collaboration where knowledge, automation, and digital reasoning systems operate together to drive innovation.

Cities Powered by Intelligent Systems

By 2040, many of the systems that support modern cities may be guided by artificial intelligence.

Urban transportation networks could rely on AI-powered coordination systems capable of optimizing traffic flows in real time. Autonomous vehicles may communicate with each other and with city infrastructure to reduce congestion and improve safety.

Energy systems may operate with greater efficiency as AI models analyze electricity demand and automatically adjust power generation across renewable and conventional sources.

Waste management, water distribution, and emergency services could also benefit from intelligent forecasting systems capable of predicting demand patterns and responding more quickly to unexpected events.

Rather than operating as isolated infrastructure networks, cities may function as integrated intelligent systems designed to improve efficiency and quality of life.

The Transformation of Work

In the intelligence economy, work may increasingly involve collaboration between humans and intelligent machines.

Many routine tasks, both physical and cognitive, may be automated by AI-driven systems. Intelligent software agents could manage administrative workflows, analyze financial data, and assist with research projects.

But this does not necessarily imply a future without human work.

Instead, the nature of work may shift toward areas where human creativity, judgment, and strategic thinking remain essential.

Professionals may increasingly rely on AI assistants that help analyze complex information, generate ideas, and simulate potential outcomes before decisions are made.

A single entrepreneur may be able to operate a company with the assistance of digital agents that manage logistics, marketing, and customer interaction.

Scientists may collaborate with AI systems capable of generating hypotheses and analyzing massive datasets.

In this sense, artificial intelligence may function less as a replacement for human workers and more as a multiplier of human capability.

Scientific Discovery at Accelerated Speed

One of the most profound effects of artificial intelligence may be its ability to accelerate scientific progress.

By 2040, machine learning systems may assist researchers in analyzing enormous datasets from fields such as genomics, climate science, and particle physics.

AI-driven simulations could help scientists test complex theories before conducting physical experiments. These simulations may dramatically reduce the time required to evaluate new technologies, materials, or medical treatments.

Drug discovery could also advance more quickly as machine learning systems analyze biological data and identify promising molecular structures.

The pace of innovation itself may accelerate.

Discoveries that once required decades of experimentation might emerge in years, or even months, through collaboration between researchers and intelligent systems.

Autonomous Infrastructure

Artificial intelligence may also transform the physical infrastructure of the global economy.

Autonomous transportation networks could manage fleets of delivery vehicles, cargo ships, and aircraft that operate with minimal human intervention.

Robotic manufacturing systems may coordinate complex production processes across global supply chains, adjusting output automatically based on real-time demand.

Agricultural systems may integrate AI-powered monitoring technologies that track soil conditions, weather patterns, and crop health, allowing farmers to optimize food production.

Energy infrastructure may increasingly rely on intelligent control systems that balance renewable energy sources with storage technologies and demand forecasting.

These developments could lead to a world where much of the physical infrastructure supporting daily life operates with a degree of autonomy.

The Expansion of Human Creativity

While automation may handle many routine tasks, human creativity may become even more valuable in the intelligence economy.

Artificial intelligence tools are already assisting artists, designers, writers, and filmmakers in generating new forms of creative expression.

By 2040, creative professionals may work alongside AI systems capable of generating visual concepts, composing music, or designing digital environments.

This collaboration may expand the range of artistic possibilities.

Rather than replacing human creativity, intelligent tools may allow individuals to experiment with ideas more rapidly and explore new creative directions.

In this sense, the intelligence economy may unlock a new era of human cultural expression.

Education in an AI-Driven World

Education systems may also evolve as artificial intelligence becomes more integrated into daily life.

Personalized learning systems could adapt educational materials to the individual needs of each student. AI tutors may assist learners in mastering complex subjects by providing guidance tailored to their pace and style of learning.

Students may also learn how to collaborate effectively with intelligent systemsdeveloping skills in critical thinking, creativity, and interdisciplinary problem solving.

In this environment, education may focus less on memorizing information and more on understanding how to apply knowledge in collaboration with intelligent tools.

The goal may shift from simply acquiring knowledge to developing the ability to think, adapt, and innovate.

Global Opportunities and Challenges

The intelligence economy may offer enormous opportunities for economic growth and scientific discovery.

But it may also introduce complex challenges.

Automation could disrupt certain job markets. Questions about data privacy, algorithmic bias, and the responsible development of advanced AI systems may become increasingly important.

Governments and institutions may need to develop new policies that balance innovation with ethical oversight.

International cooperation may also become more important as nations navigate the geopolitical implications of advanced artificial intelligence.

The technologies shaping the intelligence economy will not develop in isolation.

Their long-term impact will depend on the choices societies make about how they are designed, governed, and deployed.

A Civilization Guided by Intelligence

By 2040, artificial intelligence may be deeply integrated into the systems that support modern civilization.

Transportation networks, healthcare systems, scientific laboratories, and global supply chains may all rely on intelligent technologies that assist humans in solving complex problems.

But the most important element of the intelligence economy will remain human judgment.

Machines may analyze data and generate insights, but humans will continue to define goals, make ethical decisions, and guide the direction of technological development.

The intelligence economy is therefore not simply about machines becoming more capable.

It is about how humanity chooses to use those capabilities.

Standing at the Threshold

The world of 2040 is still being shaped.

AI is advancing rapidly, but its long-term trajectory remains uncertain. Breakthroughs in robotics, scientific discovery, and human-AI collaboration may transform industries in ways that are difficult to predict.

What is clear, however, is that intelligence, both human and artificial, will play an increasingly central role in the future of civilization.

The intelligence economy is not a distant possibility.

It is already beginning to emerge.

And the decisions made today will help determine how that future unfolds.

Key Takeaways

• By 2040, artificial intelligence may be deeply integrated into economic infrastructure and daily life.

• Cities, transportation systems, and energy networks may operate with the assistance of intelligent coordination systems.

• Human work may increasingly involve collaboration with AI-powered tools and digital agents.

• Scientific discovery may accelerate as machine learning systems assist researchers in analyzing complex datasets.

• The intelligence economy may expand opportunities for innovation, creativity, and global economic growth.

Expert Insight

Technological revolutions often unfold gradually before reaching a tipping point where their influence becomes visible across society.

Artificial intelligence may be approaching such a moment.

As intelligent systems become more capable and widely accessible, they may begin reshaping not only industries and economies but also the way humans approach knowledge, creativity, and problem solving.

The intelligence economy may therefore represent more than a technological transition.

It may represent the beginning of a new chapter in human civilization.

Strategic Question

If artificial intelligence becomes deeply integrated into the systems that support modern civilization, how should societies ensure that these technologies remain aligned with human values and priorities?

Looking Ahead

The future of artificial intelligence will not be determined solely by technological capability.

It will be shaped by the choices made by governments, businesses, researchers, and citizens around the world.

The intelligence revolution is still in its early stages.

But the decisions made during these early years may influence the trajectory of human progress for generations to come.

Chapter 22

The Economics of Intelligence

For centuries economists believed that labor, capital, and natural resources were the fundamental drivers of economic growth. Artificial intelligence introduces a new possibility: scalable intelligence.

Throughout history, economic progress has often followed advances in technology.

The agricultural revolution enabled civilizations to produce food at scale, supporting larger populations and more complex societies. The industrial revolution introduced machines capable of performing physical labor, dramatically increasing manufacturing productivity. The digital revolution accelerated the flow of information, connecting billions of people through global networks.

Artificial intelligence may represent the next transformation.

But unlike previous technological revolutions, AI is not simply a tool that enhances human effort. It is a technology that may enhance the process of thinking itself.

This shift could fundamentally reshape how economies function.

Intelligence as a Resource

In traditional economic models, growth has depended largely on three key resources: labor, capital, and natural resources.

Labor provides the human effort required to produce goods and services. Capital includes machines, infrastructure, and financial investment. Natural resources supply the raw materials necessary for production.

Artificial intelligence introduces something new.

It transforms intelligence into a scalable resource.

For centuries, intellectual work depended on the number of trained individuals available within a society. Scientific discovery, engineering innovation, and economic planning were limited by the availability of human expertise.

AI systems may change this limitation.

When intelligent software assists with analysis, research, design, and problem-solving, the effective intellectual capacity of an economy can expand.

A scientist working alongside advanced AI systems may analyze datasets that would have taken years to examine manually. Engineers may design complex systems with the help of simulation tools capable of testing thousands of possibilities in seconds.

In this way, artificial intelligence may function as a multiplier for human knowledge.

The Rise of the Intelligence Economy

As AI technologies mature, economies may begin shifting toward what could be described as an intelligence economy.

In such an economy, the most valuable resource is not simply physical labor or financial capital, but the ability to generate insights, innovations, and solutions.

Organizations capable of combining human creativity with machine intelligence may gain significant advantages.

Companies may rely on AI systems to analyze markets, design products, optimize supply chains, and accelerate research and development.

Governments may use machine learning models to better understand economic trends, environmental changes, and infrastructure needs.

Universities may integrate AI tools into scientific research, enabling faster experimentation and discovery.

Across many sectors, artificial intelligence may act as an amplifier of human capability.

Productivity and Global Growth

One of the most significant economic effects of artificial intelligence may be increased productivity.

Productivity measures how efficiently societies produce goods and services. When productivity rises, economies can generate greater output using the same or fewer resources.

Historically, major technological innovations have produced waves of productivity growth.

The steam engine powered industrial manufacturing. Electricity transformed factories and transportation systems. Computers accelerated information processing across nearly every industry.

Artificial intelligence may produce a similar effect.

By automating routine analytical tasks and assisting with complex decision-making, AI systems may allow individuals and organizations to accomplish far more than previously possible.

Some economists estimate that artificial intelligence could contribute trillions of dollars in additional economic output over the coming decades.

The long-term impact could be even greater.

If AI accelerates scientific discovery, it may lead to breakthroughs in medicine, energy systems, materials science, and transportation.

These innovations could generate entirely new industries.

New Opportunities for Entrepreneurs

Artificial intelligence may also reshape entrepreneurship.

Launching a company traditionally required significant resources. Entrepreneurs needed large teams to conduct research, develop products, and manage operations.

AI tools may reduce some of these barriers.

Small teams may be able to accomplish tasks that once required large organizations.

AI systems can assist with software development, market analysis, customer support, and operational management.

This could allow entrepreneurs to experiment with new ideas more quickly and at lower cost.

The result may be an explosion of innovation as more individuals gain the ability to build technology-driven businesses.

Just as the internet enabled a new generation of digital startups, artificial intelligence may enable the creation of entirely new categories of companies.

Challenges and Economic Transitions

While the intelligence economy offers enormous opportunities, it also introduces important challenges.

Technological revolutions often disrupt existing industries.

Automation can change the structure of employment, shifting demand for certain skills while reducing demand for others.

Artificial intelligence may produce similar effects.

Jobs that involve repetitive analytical tasks may increasingly be supported or partially automated by intelligent systems.

At the same time, new professions may emerge in fields related to AI development, system oversight, human-AI collaboration, and advanced research.

Managing this transition will require thoughtful policy and education systems that help workers adapt to changing economic conditions.

A New Phase of Economic Development

Human civilization has already experienced several major economic transformations.

Agricultural societies were defined by food production.

Industrial economies were built on manufacturing and physical infrastructure.

Information economies emerged with the rise of computers and the internet.

Artificial intelligence may signal the beginning of another phase.

An economy where the creation, amplification, and application of intelligence becomes one of the most important drivers of progress.

If developed responsibly, this transformation could lead to unprecedented levels of innovation, productivity, and scientific discovery.

But the direction it takes will depend on the choices societies make in the coming years.

The intelligence economy is only beginning to emerge.

And its future is still being written.

Key Takeaways

• Artificial intelligence represents one of the most significant technological transformations in human history.

• Machine intelligence has the potential to reshape economies, industries, and scientific discovery.

• The future of AI will be shaped not only by technology but also by policy, ethics, and global cooperation.

• Human creativity, leadership, and responsibility will remain essential in guiding the development of AI.

• The decisions made today will influence how artificial intelligence shapes the future of civilization.

Expert Insight

Throughout history, transformative technologies have often expanded humanity's capabilities while introducing new challenges. Artificial intelligence may represent the most powerful example of this pattern.

Machine intelligence could accelerate scientific discovery, improve healthcare, optimize global infrastructure, and expand access to knowledge. At the same time, it raises questions about governance, economic inequality, and the responsible development of powerful technologies.

The trajectory of artificial intelligence will therefore depend not only on engineers and researchers, but also on policymakers, educators, entrepreneurs, and citizens.

The future of machine intelligence is ultimately a human decision.

Strategic Question

If humanity is creating a new form of intelligence, what responsibilities come with that power?

How can societies ensure that artificial intelligence advances human progress while preserving the values that define civilization?

Looking Ahead

AI is still in its early stages.

The technologies described throughout this book will continue to evolve in ways that are difficult to predict. New discoveries, new industries, and new challenges will emerge as machine intelligence becomes more deeply integrated into human society.

What remains certain is that the choices made today will shape the direction of this revolution.

The next intelligence revolution has already begun.

The future it creates will depend on how we choose to guide it.

Chapter 23

Five Forces Driving the Intelligence Revolution

Every technological revolution has a structure. The intelligence economy will have its own stack of technologies, platforms, and applications.

Understanding the Transformation

Throughout this book, we have explored how artificial intelligence is reshaping technology, industries, and the global economy.

Yet technological revolutions are rarely driven by a single innovation. They are usually the result of several forces converging at the same moment in history.

The industrial revolution was powered by the combination of steam engines, factory systems, and expanding transportation networks. The digital revolution emerged from the convergence of computers, telecommunications infrastructure, and the internet.

The intelligence revolution follows a similar pattern.

AI is advancing rapidly because several powerful forces are accelerating progress simultaneously. Understanding these forces helps explain why the current transformation is unfolding so quickly, and why its impact may be so profound.

Five forces, in particular, are shaping the rise of the intelligence economy.

Force One: Computing Power

The first driving force behind the intelligence revolution is computing power.

Modern artificial intelligence systems require enormous amounts of computation. Training advanced machine learning models involves performing trillions of mathematical operations across vast datasets.

Over the past two decades, advances in semiconductor technology have dramatically increased the amount of computing power available to researchers and organizations.

Specialized processors designed for machine learning workloads, such as graphics processing units and dedicated AI chips, allow complex neural networks to be trained more efficiently.

At the same time, global cloud computing platforms provide access to powerful computing infrastructure that can scale on demand.

These developments have created the computational foundation required to support modern artificial intelligence.

Force Two: Data

Artificial intelligence systems learn by analyzing data.

The rapid growth of the internet, mobile devices, and digital services has produced an unprecedented quantity of information in machine-readable form.

Text, images, financial records, scientific research, and human communication are now stored and transmitted through digital networks.

This massive expansion of data has provided machine learning systems with the training material needed to recognize patterns and generate predictions.

The availability of large datasets has therefore played a crucial role in enabling recent breakthroughs in artificial intelligence.

Force Three: Algorithms

While computing power and data provide the raw resources for AI development, algorithms determine how those resources are used.

Advances in machine learning techniques, particularly deep learning and transformer architectures, have dramatically improved the ability of computers to analyze complex information.

These algorithms allow AI systems to process language, recognize images, generate software code, and identify patterns within enormous datasets.

As researchers continue refining these methods, the capabilities of artificial intelligence systems continue to expand.

Force Four: Global Investment

Artificial intelligence has become one of the most heavily funded technological fields in modern history.

Governments, corporations, and venture capital investors are investing billions of dollars into AI research, infrastructure, and startup development.

Major technology companies are constructing large data centers capable of training increasingly powerful AI systems.

Meanwhile, entrepreneurs around the world are building companies that apply machine learning to industries such as healthcare, finance, manufacturing, and logistics.

This surge of investment has accelerated the pace of innovation across the entire AI ecosystem.

Force Five: Human Curiosity

The final, and perhaps most important, force behind the intelligence revolution is human curiosity.

Throughout history, scientific discovery and technological progress have been driven by the desire to understand the world and explore new possibilities.

Artificial intelligence research reflects this same impulse.

Scientists seek to understand how learning systems operate. Engineers experiment with new ways of designing intelligent machines. Entrepreneurs explore how AI technologies might solve real-world problems.

This collective curiosity continues to push the boundaries of what artificial intelligence can achieve.

As long as this drive for discovery persists, the development of intelligent technologies will continue to advance.

The Convergence of Forces

Individually, each of these forces would represent an important technological development.

But together, they create something far more powerful.

The convergence of computing power, data availability, algorithmic innovation, global investment, and human curiosity has produced the conditions necessary for rapid progress in artificial intelligence.

This convergence explains why AI capabilities have advanced so dramatically during the past decade.

It also suggests that the intelligence revolution may continue accelerating in the years ahead.

Looking Toward the Future

Understanding the forces behind the intelligence revolution helps illuminate the path forward.

As computing infrastructure expands, new algorithms are developed, and global investment continues to grow, artificial intelligence systems will likely become more capable.

These advances may reshape industries, accelerate scientific discovery, and influence how societies organize economic activity.

But technology alone does not determine the future.

Human values, governance decisions, and collective priorities will influence how these capabilities are applied.

The intelligence revolution therefore represents both a technological transformation and a societal choice.

The chapters that follow explore how humanity may guide this transformation responsibly as machine intelligence becomes an increasingly powerful partner in human progress.

Key Takeaways

• Artificial intelligence is no longer confined to isolated tools or experimental technologies; it is becoming a foundational layer of modern economic and social systems.

• The integration of intelligent systems across industries is accelerating the pace of innovation, productivity, and global competition.

• As AI capabilities expand, the implications extend beyond technology to questions of governance, ethics, economic distribution, and long-term societal stability.

• The intelligence revolution will influence not only how industries operate, but also how opportunity, power, and decision-making are structured in the future economy.

• Understanding the broader implications of intelligent systems is essential for leaders, policymakers, and individuals navigating the next phase of technological transformation.

Expert Insight

Technological revolutions rarely unfold in isolation. They reshape institutions, economic incentives, and social structures in ways that are often difficult to predict in the early stages of change.

The intelligence revolution is no exception. As AI systems become more capable and more widely integrated into global infrastructure, the decisions made today about their development, governance, and application will influence the trajectory of societies for decades to come.

The challenge is not simply building more advanced systems. It is ensuring that the frameworks guiding their use evolve alongside the technology itself.

Strategic Question

If intelligence becomes a scalable technological resource, how should societies ensure that the benefits of this transformation are distributed in ways that promote long-term prosperity, stability, and opportunity?

Looking Ahead

The rapid progress of artificial intelligence raises a deeper question that extends beyond technology itself.

What principles should guide the intelligence revolution as it reshapes industries, economies, and the structure of global power?

In the next chapter, we step back from individual technologies and examine the broader philosophical and strategic foundations that may shape the future of the intelligence age.

PART VII

THE BIG QUESTIONS

The Big Idea

The intelligence revolution is not just a technological transformation; it is a civilizational turning point.

For the first time in human history, intelligence itself is becoming scalable. Machines are learning, reasoning, creating, and solving problems at speeds and scales that were previously impossible.

This shift will reshape industries, redefine economic power, and transform the structure of opportunity around the world.

But the future of the intelligence age will not be determined by technology alone.

It will be determined by the choices societies make about how intelligence is developed, deployed, and governed.

The most important questions of this revolution are no longer about what machines can do.

They are about what humanity chooses to do next.

Why This Matters Now

For most of human history, technological revolutions unfolded slowly enough that societies could adapt over generations. The intelligence revolution is different.

Artificial intelligence is advancing at a pace that compresses decades of change into years, and sometimes into months. Entire industries are being redesigned, new economic structures are emerging, and the nature of human work is being redefined in real time.

Yet the most important questions are no longer purely technical.

They are human.

Who will control the systems that generate intelligence at scale?

How will wealth created by intelligent machines be distributed?

What happens to work, creativity, and identity when machines can think, learn, and solve problems alongside us?

These are not distant philosophical debates. They are decisions being made right now by governments, companies, and individuals.

The answers will determine whether the intelligence revolution leads to unprecedented prosperity, or deep structural imbalance.

Part VII explores the questions that will shape that outcome.

Before we can decide what kind of future we want to build, we must first confront the questions that define it.

Chapter 24

The Intelligence Revolution Manifesto

Humanity has entered a new era, one in which intelligence itself is becoming a technology.

The Most Important Technology of Our Time

Every generation experiences technological change.

But occasionally, humanity encounters a transformation so powerful that it reshapes the trajectory of civilization itself.

The printing press expanded access to knowledge and helped ignite the scientific revolution. The steam engine powered industrial societies and transformed global commerce. Electricity reshaped cities and industry. The internet connected billions of people across a global information network.

Artificial intelligence may represent the next transformation of this magnitude.

For the first time in the history of Earth, intelligence itself is becoming a scalable technology.

Machines are beginning to analyze information, generate ideas, and assist with complex decision-making. These capabilities may accelerate scientific discovery, reshape industries, and influence nearly every aspect of human life.

We are witnessing the early stages of the intelligence revolution.

Technology Reflects Human Choices

Despite the extraordinary capabilities of artificial intelligence, technology alone does not determine the future.

Machines do not set goals.

Humans do.

Every technological system reflects the intentions of the people who design it. The structure of our economies, the priorities of our institutions, and the values embedded within our technologies will shape the direction of the intelligence revolution.

Artificial intelligence can amplify creativity, accelerate discovery, and help solve global challenges.

But it can also magnify inequality, concentrate power, or create unintended consequences if developed without careful consideration.

The future of AI will therefore depend not only on algorithms but on human judgment.

The Responsibility of Builders

Throughout history, the people who built new technologies often carried enormous influence over the direction of society.

Engineers designed the machines that powered factories. Scientists developed the technologies that enabled space exploration. Computer pioneers built the digital networks that now connect the world.

Today, a new generation of builders is emerging.

Researchers developing machine learning systems. Engineers constructing global computing infrastructure. Entrepreneurs launching companies powered by artificial intelligence.

These individuals are not simply building products.

They are helping shape the technological architecture of the future.

With that influence comes responsibility.

The decisions made by today's builders may influence how intelligent technologies interact with society for decades to come.

The Opportunity Before Us

The intelligence revolution is not only about risk or disruption.

It is also about possibility.

Artificial intelligence may help accelerate medical research and develop treatments for diseases that have long challenged humanity.

Machine learning systems may assist scientists in understanding climate systems, designing new materials, and discovering sustainable energy technologies.

Intelligent tools may help individuals learn more effectively, create new forms of art, and solve complex problems.

In the best possible future, artificial intelligence becomes a partner in human progress.

A technology that expands what individuals and societies are capable of achieving.

The Role of Human Intelligence

Despite the rapid progress of machine intelligence, human capabilities remain central to the future of civilization.

Creativity, empathy, ethical judgment, and long-term vision are qualities that cannot easily be reduced to algorithms.

Human societies will continue to determine the goals that technology pursues.

Artificial intelligence may generate insights and assist with analysis, but the direction of progress will ultimately be guided by human values.

In this sense, the intelligence revolution is not about replacing human intelligence.

It is about augmenting it.

A Global Conversation

The rise of artificial intelligence is not confined to laboratories or technology companies.

It is a global development that affects governments, businesses, educators, researchers, and citizens around the world.

The decisions made about how AI systems are developed, regulated, and deployed will shape the future of societies everywhere.

This requires open dialogue.

Nations must collaborate on shared principles for responsible AI development. Institutions must consider the ethical implications of increasingly powerful technologies.

And individuals must remain engaged in the conversation about how these systems influence the world they inhabit.

Standing at the Beginning

It is important to remember that humanity is still at the beginning of the intelligence revolution.

AI is advancing rapidly, but many of the technologies that will define this era are still emerging.

The next several decades may bring breakthroughs in robotics, scientific discovery, and human-AI collaboration that are difficult to imagine today.

The intelligence revolution will not unfold overnight.

But its influence may shape the future of civilization for generations.

The Future We Choose

Technology expands what is possible.

But it does not determine what is inevitable.

The future of artificial intelligence will be shaped by the decisions humanity makes about how to develop and use these powerful systems.

We can choose to build technologies that amplify human creativity, strengthen global cooperation, and accelerate scientific discovery.

Or we can allow innovation to proceed without careful consideration of its broader consequences.

The intelligence revolution is not merely a technological story.

It is a human one.

And the next chapter of that story is still being written.

Key Principles of the Intelligence Revolution

• Intelligence is becoming a scalable technology.

• Artificial intelligence can amplify human creativity and productivity.

• The direction of technological progress is shaped by human values and decisions.

• Responsible development of AI requires global collaboration and thoughtful governance.

• The intelligence revolution represents one of the most significant transformations in human history.

Final Reflection

The emergence of machine intelligence marks the beginning of a new technological era.

Humanity now possesses the ability to create systems that can assist with reasoning, discovery, and innovation at extraordinary scale.

How we choose to use this capability may influence the future of civilization itself.

The intelligence revolution has begun.

The future now depends on the choices we make.

Chapter 25

What This Means for You

Technological revolutions reshape industries, economies, and societies; they also reshape individual opportunity. If artificial intelligence truly becomes a scalable resource, the size of the global economy could expand dramatically.

The intelligence revolution is not limited to governments, research laboratories, or technology companies.

It affects individuals as well.

Students entering the workforce will collaborate with intelligent tools that amplify their abilities. Entrepreneurs may build companies using AI systems that accelerate product development and market analysis. Researchers may rely on machine learning to discover new medicines, materials, and technologies.

Understanding artificial intelligence is therefore becoming an essential form of literacy.

Those who learn how to work effectively with intelligent systems may gain extraordinary advantages in the decades ahead.

But perhaps the most important question is not technological.

It is philosophical.

What kind of future do we want to build with these powerful tools?

Artificial intelligence expands what humanity is capable of achieving.

The challenge, and the opportunity, lies in ensuring that these capabilities are used to advance knowledge, improve quality of life, and expand the possibilities of human progress.

The intelligence revolution has begun.

And every generation now has a role in shaping what comes next.

The Scale Question

AI is advancing at a remarkable pace.

Machines can now write software, analyze scientific data, assist with medical diagnosis, and perform complex reasoning tasks that once required highly trained professionals.

But these developments raise a deeper question.

A question that extends far beyond technology.

What happens when intelligence itself becomes scalable?

For most of human history, intelligence has been limited by biology. A society's capacity to solve problems depended on the number of skilled individuals it could educate and train.

Artificial intelligence changes that equation.

If machines can assist with reasoning, research, and decision-making, the effective intellectual capacity of society could increase dramatically.

A scientist working alongside advanced AI systems may be able to analyze data faster, test more hypotheses, and explore more ideas than ever before.

Engineers may design new materials and technologies with the help of machine learning models capable of simulating millions of possibilities.

Entrepreneurs may build companies with small teams supported by intelligent software capable of automating large portions of business operations.

In this sense, artificial intelligence may act as a multiplier for human creativity and problem-solving.

But the implications extend further.

If intelligence can be expanded through machines, the rate of scientific discovery may accelerate.

New medicines could be developed more quickly.

Energy technologies could become more efficient.

Engineering breakthroughs could emerge in fields ranging from transportation to climate science.

Some economists have suggested that artificial intelligence could eventually add tens of trillions of dollars to the global economy.

Others believe its impact may be even larger.

But these possibilities depend on how the technology is developed and governed.

AI is not a force that unfolds automatically.

It is shaped by the decisions made by researchers, companies, governments, and societies.

The same technologies that accelerate discovery can also introduce new risks.

AI systems must be designed responsibly.

Their use must be guided by ethical frameworks and thoughtful policy.

And their benefits must be distributed broadly across society.

Humanity now stands at a turning point.

For the first time at global scale, we are beginning to create machines capable of assisting with the process of thinking.

How we choose to develop this technology may influence the trajectory of civilization for generations.

Artificial intelligence could become one of the most powerful tools ever created for expanding human potential.

Or it could deepen existing inequalities and create new challenges that societies must confront.

The outcome is not predetermined.

It will depend on the choices we make in the years ahead.

And those choices begin now.

Chapter 26

Big Ideas from the Intelligence Revolution

The intelligence revolution is not just a technological shift. It is a transformation of how humanity creates knowledge, wealth, and progress.

Throughout this book, we have explored the technologies, industries, and global forces shaping the rise of artificial intelligence.

While the details of these developments are complex, several fundamental ideas emerge that help explain why artificial intelligence may become one of the most important transformations in human history.

These ideas provide a framework for understanding the intelligence revolution now unfolding.

Intelligence Is Becoming Scalable

For most of human history, intelligence was limited by biology.

Human reasoning, creativity, and learning emerged from the structure of the brain, and the growth of knowledge depended on human effort. Scientific progress, technological innovation, and economic development were all constrained by the pace at which people could think, analyze, and discover.

Artificial intelligence changes this equation.

Machine learning systems allow certain forms of reasoning, pattern recognition, and analysis to operate at digital scale. Instead of relying solely on human cognition, societies can now amplify intellectual work through computational systems.

This shift may dramatically accelerate discovery across many fields.

Data and Computing Power Are the Engines of AI

The rapid progress of artificial intelligence is not the result of a single breakthrough.

It is the convergence of several technological forces.

Large datasets provide the raw material for machine learning. Advanced algorithms enable systems to learn from that data. Massive computing infrastructure allows these models to process enormous quantities of information.

Together, these elements form the technological engine of modern artificial intelligence.

As computing power continues to expand and new algorithms are developed, the capabilities of AI systems may continue to grow.

Artificial Intelligence Is Becoming a Platform Technology

Some technologies influence only a narrow set of applications.

Others become platforms that support entire ecosystems of innovation.

Electricity became a platform for industrial machinery and household appliances. The internet became a platform for communication, commerce, and digital media.

Artificial intelligence may become a similar platform technology.

Machine learning systems are now being applied to healthcare, finance, transportation, scientific research, and countless other fields.

The result is a broad technological transformation that extends far beyond computer science.

The AI Revolution Is Global

AI is not being developed in isolation.

Governments, universities, technology companies, and research institutions around the world are investing heavily in AI development.

The United States, China, Europe, and many other regions are competing to build the infrastructure, talent, and research ecosystems required to lead in artificial intelligence.

This global competition may influence economic growth, technological leadership, and geopolitical power for decades to come.

Human Intelligence Still Matters

Despite the extraordinary capabilities of machine learning systems, artificial intelligence does not replace human intelligence.

It extends it.

Human creativity, ethical judgment, leadership, and imagination remain essential for guiding the development of technology.

Artificial intelligence can analyze data and generate insights, but humans determine how those insights are used.

The future of AI will therefore depend not only on technological progress, but also on human choices.

Enduring Laws and Temporary Manifestations

One of the most useful distinctions a reader can carry forward from this book is the difference between enduring laws and temporary manifestations.

Enduring laws are the deep principles that will remain true regardless of which companies, architectures, or nations lead at any given moment. Temporary manifestations are the specific actors, tools, and configurations that embody those principles right now and which will change as the technology continues to evolve.

For example:

Enduring law: Intelligence becomes an economic force multiplier. Whatever form it takes, scalable intelligence amplifies the productive capacity of individuals, organizations, and economies.

Temporary manifestation: The specific model architectures, hardware manufacturers, and national leaders that dominate AI development today. These will change some within years, some within decades.

Enduring law: Control over intelligence infrastructure confers strategic advantage. Whether that infrastructure is a clay tablet, a printing press, a telephone network, or a computing cluster, the pattern repeats.

Temporary manifestation: Semiconductor export controls, specific cloud providers, the US-China competition as it stands today. These are the current expressions of that enduring pattern.

Enduring law: Transitions between technological eras create periods of institutional disruption before new equilibria form. The disruption is temporary; the structural change it produces is not.

Temporary manifestation: Which professions are being disrupted right now, which companies are leading, and which regulatory frameworks are being debated.

Readers who internalize this distinction will find that this book remains useful long after the specific names and numbers it contains have become historical footnotes. The laws will not change. The manifestations will.

The Intelligence Revolution Is Only Beginning

Perhaps the most important idea explored in this book is that artificial intelligence is still in its early stages.

Many of the systems we see today represent only the first generation of machine intelligence.

New research, new computing architectures, and new applications will likely expand the capabilities of AI far beyond what currently exists.

Just as the early internet of the 1990s eventually evolved into today's global digital ecosystem, artificial intelligence may continue to evolve for decades.

The intelligence revolution has begun.

But its ultimate impact remains unwritten.

Chapter 27

A Framework for Reading AI Beyond This Book

The next intelligence shift is already being built. This chapter gives you the tools to recognize it before it arrives.

The Problem with Watching the Present

Most people discover technological revolutions after they have already happened. They read about the steam engine after the factories are built. They encounter the internet after the networks are already running. They engage with artificial intelligence after it has already entered the economy.

This chapter offers a different approach. Rather than describing what exists today, it provides a framework for identifying intelligence shifts before they become obvious and for evaluating claims about AI with clear, durable criteria.

Signal One: Watch Where Friction Disappears

Every intelligence shift reduces friction somewhere. The printing press removed the friction of hand-copying manuscripts. The telegraph removed the friction of distance in communication. The internet removed the friction of information access.

When you encounter a new AI capability, ask: what task that used to require significant time, expertise, or infrastructure can now be accomplished more easily? If the answer is significant if the friction reduction is large enough to change behavior at scale you are watching the early stages of a shift worth following.

Signal Two: Watch Where Infrastructure Investment Concentrates

Major technological transitions require enormous physical investment before they become visible to most people. Railroad tracks were laid for years before ordinary citizens took train journeys. Electrical infrastructure was built for decades before homes were reliably lit.

When governments and corporations begin committing capital at extraordinary scale to a new category of infrastructure power generation, computing, connectivity, logistics that investment is a reliable leading indicator of transformation. Follow the infrastructure, and you will find the next shift.

Signal Three: Watch Where Scientific Disciplines Converge

The most transformative technologies emerge when separate fields of knowledge unexpectedly solve each other's problems. The steam engine emerged from the intersection of metallurgy, thermodynamics, and industrial design. Modern AI emerged from the convergence of neuroscience-inspired architectures, mathematical optimization, and computing hardware originally built for an entirely different purpose.

Watch for disciplines that rarely communicate with each other beginning to publish collaborative research. Watch for engineers solving problems by borrowing methods from biology, chemistry, or physics. These convergences are early signals of incoming transformation.

Signal Four: Watch What Becomes Cheap

Technologies that change civilization tend to make something dramatically cheaper. Printing made knowledge cheap. Electricity made mechanical power cheap. The internet made communication and information retrieval cheap.

AI is making certain forms of reasoning, analysis, and content generation cheap. The next intelligence shift will make something else cheap something that is currently expensive, specialized, or inaccessible. When you see a capability that was once rare becoming abundant and affordable, pay close attention. That is the early signature of transformation.

Signal Five: Watch How Resistance Forms

Every genuine transformation encounters organized resistance from those whose position depends on the existing order. This resistance is not merely conservatism it is often well-reasoned and sometimes correct in its concerns. But the form resistance takes tells you something important about the nature of the shift.

If resistance is focused on safety and governance on how to deploy the technology responsibly rather than whether to stop it the shift is likely already inevitable and the real question has moved to terms of adoption. If resistance is focused on denial or reversibility arguing that the technology will not work or will be abandoned the shift is likely still early enough that its direction remains uncertain.

How to Evaluate Any AI Claim

Whether you are reading a news article, evaluating an investment, or deciding whether to adopt a new tool, the following questions provide a reliable filter for any AI-related claim:

Is this an enduring shift or a temporary manifestation? Ask whether the underlying capability is genuinely new, or whether a specific company or product is simply the current face of a deeper change that will persist beyond any one actor.

What friction is being removed? If a capability cannot clearly answer this question, it may be a novelty rather than a transformation. Real shifts remove friction that was previously accepted as inevitable.

Who is building infrastructure around it? Hype creates commentary. Genuine shifts attract infrastructure investment. Look not at what people are saying, but at what they are building.

What would have to be true for this to fail? The most useful test of any prediction is to ask what conditions would need to hold for it to be wrong. This clarifies whether confidence in a claim is justified or whether it is being asserted without meaningful constraint.

A Note on Humility

No framework predicts the future with certainty. The signals described here are indicators, not guarantees. The most important habit any reader can develop is the willingness to update their view as new evidence arrives.

This book was written during a particular moment in the intelligence revolution when certain technologies, certain companies, and certain geopolitical configurations were dominant. Those specifics will change. But the dynamics described here convergence, threshold crossing, infrastructure buildout, friction reduction, and the enduring relationship between intelligence and economic power will repeat.

The reader who finishes this book knowing only what exists today has learned history. The reader who finishes it knowing how to recognize what comes next has learned something far more durable.

A Final Thought

Every generation lives through moments that redefine what is possible.

The agricultural revolution reshaped how humans lived.

The industrial revolution reshaped how humans worked.

The digital revolution reshaped how humans connect and share knowledge.

Now the intelligence revolution is reshaping how humans create, discover, and solve problems.

The technologies emerging today will influence economies, industries, and societies for decades to come. But technology alone does not determine the future.

The systems we build reflect human choices: our values, our priorities, and how we decide to use the tools we create.

Artificial intelligence will expand what humanity can achieve.

What humanity chooses to do with that power will define the world that follows.

The intelligence revolution is only beginning.

And the future it creates is still ours to shape.

Epilogue

The Future We Choose

Artificial intelligence represents one of the most powerful technologies humanity has ever created.

For the first time in the history of our planet, intelligence is no longer limited to biological organisms. Machines are beginning to learn, reason, and assist with complex problem-solving.

This transformation is still in its early stages.

The systems that exist today represent only the first generation of machine intelligence. Over the coming decades, advances in computing infrastructure, scientific research, and algorithm design may produce technologies far more capable than those we see today.

Artificial intelligence could accelerate scientific discovery, transform global industries, and expand human productivity on a scale never before experienced.

But technology alone does not determine the future.

The direction of the intelligence revolution will depend on the choices made by people.

Researchers will decide how systems are designed.

Entrepreneurs will decide how they are deployed.

Governments will decide how they are regulated.

And societies will decide how their benefits are shared.

Throughout history, technological revolutions have reshaped civilization.

The printing press expanded knowledge.

The industrial revolution transformed production.

The internet connected the world.

Artificial intelligence may represent the next chapter in that story.

For the first time at global scale, humanity is creating systems capable of performing forms of reasoning and problem-solving once limited to biological minds.

This moment represents more than a technological transition.

It represents the beginning of a new chapter in the story of intelligence on Earth.

Whether artificial intelligence becomes one of humanity's greatest achievements, or one of its greatest challenges, will depend on how wisely we guide its development.

The intelligence revolution has begun.

The future of artificial intelligence has not yet been written.

And the world that follows will be the one we choose to build.

—Saint Blanc

Discussion & Reflection Guide

The following questions are designed to encourage deeper reflection on the themes explored throughout this book. AI is not only a technological development but also a societal transformation that raises important questions about economics, ethics, governance, and the future of human progress.

These questions may be useful for classroom discussions, professional workshops, or individual reflection.

Understanding the Rise of Artificial Intelligence

What technological breakthroughs made modern artificial intelligence possible, and why did these developments accelerate so rapidly in recent years?

How did the convergence of data availability, computing power, and machine learning algorithms enable the AI revolution?

Why did artificial intelligence remain largely confined to research laboratories for decades before becoming widely adopted?

The Global AI Competition

Why are governments increasingly viewing artificial intelligence as a strategic technology?

How might AI influence geopolitical power and economic leadership in the twenty-first century?

What role do semiconductor manufacturing and computing infrastructure play in the global AI race?

AI and the Transformation of Work

How might artificial intelligence change the nature of work across different industries?

Which professions are most likely to be affected first by AI-driven automation?

What skills may become most valuable in an economy where humans collaborate with intelligent machines?

Artificial Intelligence Across Industries

How is artificial intelligence already transforming industries such as healthcare, finance, and robotics?

What new opportunities might emerge as AI technologies become more widely accessible?

Which industries may experience the most dramatic changes over the next two decades?

Ethics and Responsibility

What ethical challenges arise as artificial intelligence systems become more powerful?

How can societies ensure that AI technologies are developed responsibly and used in ways that benefit humanity?

Who should be responsible for establishing guidelines and governance frameworks for advanced AI systems?

The Future of Intelligence

What might the world look like if artificial intelligence continues advancing at its current pace?

How might human creativity, leadership, and decision-making evolve in a world where intelligent machines are widely available?

What responsibilities do humans carry as creators of increasingly powerful intelligent systems?

The Next Intelligence Revolution

What do you believe will be the most significant impact of artificial intelligence on society over the next twenty years?

How should individuals, organizations, and governments prepare for the technological transformations described in this book?

Acknowledgments

The creation of any book is rarely the work of one individual alone. It is the result of countless ideas, conversations, research efforts, and technological advancements developed by many people across generations.

First and foremost, I would like to acknowledge the scientists, researchers, engineers, and thinkers who have contributed to the development of artificial intelligence. Their decades of work in computer science, mathematics, neuroscience, and engineering laid the foundation for the technologies discussed throughout this book.

The breakthroughs in machine learning, neural networks, and computational theory described in these pages are the result of collective efforts by researchers at universities, laboratories, and technology companies around the world.

I would also like to recognize the authors, analysts, and journalists whose work has helped explain the rapid evolution of artificial intelligence to a wider audience. Their insights into technology, economics, and policy have been invaluable in shaping the broader understanding of this field.

Finally, I extend appreciation to the readers who take the time to explore the ideas presented here. AI is not merely a technical subject; it is a societal transformation that will influence economies, industries, and daily life for generations to come. Public understanding and thoughtful discussion will play an important role in guiding how these technologies develop.

The future of artificial intelligence will ultimately be shaped by the decisions made by people across many disciplines: scientists, entrepreneurs, policymakers, educators, and citizens.

It is my hope that this book contributes, in some small way, to a deeper understanding of the opportunities and responsibilities that accompany the rise of machine intelligence.

—Saint Blanc

References and Research Sources

AI is a rapidly evolving field drawing on research from computer science, economics, public policy, and technology development. The following books, reports, and research papers informed the ideas and analysis discussed throughout this work.

Foundational Artificial Intelligence Research

Alan Turing (1950). Computing Machinery and Intelligence. Mind Journal.

Geoffrey Hinton, Yann LeCun, & Yoshua Bengio. Research contributions on deep neural networks and machine learning.

OpenAI. GPT-4 Technical Report.

Google DeepMind. Publications on large-scale machine learning systems.

Artificial Intelligence and Society

Eric Schmidt, Henry Kissinger, & Daniel Huttenlocher. The Age of AI: And Our Human Future.

Mustafa Suleyman. The Coming Wave: Technology, Power, and the Twenty-First Century's Greatest Dilemma.

Nick Bostrom. Superintelligence: Paths, Dangers, Strategies.

Artificial Intelligence and the Global Economy

McKinsey Global Institute. The Economic Potential of Artificial Intelligence.

PwC. Sizing the Prize: What's the Real Value of AI for Your Business?

Stanford University. AI Index Report.

Technology Infrastructure and Computing

NVIDIA. Research and technical publications on GPU computing for artificial intelligence.

MIT. Computer Science and Artificial Intelligence Laboratory (CSAIL) publications.

IBM Research. Research papers on machine learning and cloud computing infrastructure.

Artificial Intelligence and Innovation

Kai-Fu Lee. AI 2041: Ten Visions for Our Future.

Andrew Ng. Publications on machine learning systems and AI entrepreneurship.

World Economic Forum. Reports on artificial intelligence and the future of work.

Further Reading

Readers interested in exploring the subjects discussed in this book in greater depth may wish to consult the following works:

Artificial Intelligence: A Modern Approach

Life 3.0

The Second Machine Age

Machine Learning Yearning

Data Sources

Stanford AI Index

OECD AI Policy Observatory

United Nations technology and development reports

These publications regularly cover new developments in machine learning research, robotics, artificial intelligence policy, and emerging technology trends.

Closing Note

AI is a rapidly evolving field. New discoveries, improved algorithms, and expanding computing infrastructure continue to reshape the capabilities of machine intelligence. The sources referenced here represent only a portion of the global research landscape informing the ongoing development of artificial intelligence technologies.

Continue the Journey

The Intelligence Revolution has only just begun.

Artificial intelligence is not simply transforming technology; it is reshaping the global economy, redefining the nature of work, and creating entirely new systems of power and wealth.

In the next book of the series, we explore what happens when artificial intelligence begins to reshape the economic foundations of society.

Coming Next

Book 2 The Intelligence Economy

From automated financial systems to AI-driven corporations and intelligent supply chains, the next phase of the intelligence revolution will transform how the world creates and distributes value.

The question is no longer whether AI will influence the economy.

The question is who will control it.

About the Author

Saint Blanc is an entrepreneur, technology thinker, and U.S. Navy veteran. After serving as an Electrician's Mate in the United States Navy, he attended Jacksonville University, where he earned a Bachelor of Science in Sociology with minors in Chemistry and Biology.

Blanc later became an entrepreneur and the founder of InExt Home, pursuing ventures focused on innovation, technology, and future-driven ideas.

The Next Intelligence Revolution is his first book in a series exploring how artificial intelligence will reshape economies, industries, and society.